华章IT
HZBOOKS | Information Technology

智能系统与技术丛书

Keras Deep Learning Cookbook

Keras深度学习实战

[印] 拉蒂普·杜瓦（Rajdeep Dua） 著
曼普里特·辛格·古特（Manpreet Singh Ghotra）

罗娜 祁佳康 译

机械工业出版社
China Machine Press

图书在版编目（CIP）数据

Keras 深度学习实战 /（印）拉蒂普·杜瓦（Rajdeep Dua）等著；罗娜，祁佳康译．—北京：机械工业出版社，2019.5

（智能系统与技术丛书）

书名原文：Keras Deep Learning Cookbook

ISBN 978-7-111-62627-5

I. K… II. ①拉… ②罗… ③祁… III. 软件设计 IV. TP311.5

中国版本图书馆 CIP 数据核字（2019）第 084161 号

本书版权登记号：图字 01-2018-8351

Rajdeep Dua, Manpreet Singh Ghotra: Keras Deep Learning Cookbook (ISBN: 978-1-78862-175-5).

Copyright © 2018 Packt Publishing. First published in the English language under the title "Keras Deep Learning Cookbook".

All rights reserved.

Chinese simplified language edition published by China Machine Press.

Copyright © 2019 by China Machine Press.

本书中文简体字版由 Packt Publishing 授权机械工业出版社独家出版。未经出版者书面许可，不得以任何方式复制或抄袭本书内容。

Keras 深度学习实战

出版发行：机械工业出版社（北京市西城区百万庄大街22号 邮政编码：100037）			
责任编辑：刘　锋		责任校对：殷　虹	
印　　刷：北京文昌阁彩色印刷有限责任公司		版　次：2019年6月第1版第1次印刷	
开　　本：186mm × 240mm　1/16		印　张：12.5	
书　　号：ISBN 978-7-111-62627-5		定　价：69.00元	

凡购本书，如有缺页、倒页、脱页，由本社发行部调换

客服热线：（010）88379426　88361066　　　投稿热线：（010）88379604

购书热线：（010）68326294　　　　　　　　　读者信箱：hzit@hzbook.com

版权所有·侵权必究

封底无防伪标均为盗版

本书法律顾问：北京大成律师事务所　韩光／邹晓东

THE TRANSLATOR'S WORDS
译 者 序

作为机器学习中的一个重要研究领域，人工神经网络的发展历史一波三折。2006年以来，随着深度学习的兴起和成功应用，人工神经网络迎来了新的生机。与传统的人工神经网络相比，深度学习的最大特点是网络层数有了大幅度的增加，配合其他相关技术，解决了以图像、声音等作为数据的传统人工神经网络难以解决的问题。作为快速实现深度学习的平台，TensorFlow一定程度上简化了神经网络的构建程序，Keras进一步对TensorFlow进行了封装，从而能更加快速地把用户的思想转化为代码。

本书从实用的角度出发，全方面介绍了如何使用Keras解决深度学习中的各类问题。本书假设读者无任何关于深度学习编程的基础知识，首先介绍了Keras这一高度模块化、极简式的深度学习框架的安装、配置和编译等平台搭建知识，而后详细介绍了深度学习所要求的数据预处理等基本内容，在此基础上介绍了卷积神经网络、生成式对抗网络、递归神经网络这三种深度学习方法并给出了相关实例代码，最后本书介绍了自然语言处理、强化学习两方面的内容。

本书是一本实践性很强的深度学习工具书，既适合希望快速学习和使用Keras深度学习框架的工程师、学者和从业者，又特别适合立志从事深度学习和AI相关的行业并且希望用Keras开发实际项目的工程技术人员。

本书翻译工作得到国家自然科学基金项目（项目编号：61403140）的资助，在此表示衷心感谢。

感谢华章公司的刘锋编辑不辞辛苦地与译者沟通相关细节内容，同时感谢他在翻译本书过程中给予的诸多帮助。

限于本人水平，难免会对本书中部分内容的理解或中文语言表达存在不当之处，敬请读者批评指正，以便能够不断改进。

罗娜　祁佳康
2019年于上海

ABOUT THE REVIEWER
审校者简介

  Sujit Pal 工作于 Reed-Elsevier PLC Group 内的 Elsevier 实验室，研究涉及信息检索、分布式处理、本体开发、自然语言处理、机器学习，并用 Python、Scala 和 Java 进行开发。通过结合在这些领域的专长，帮助公司实现产品功能构建或改进。他相信终身学习，并长期在 sujitpal.blogspot.com 上记录他的一些见解。

PREFACE

前　　言

Keras 采用 Python 编写，能够快速准确地训练卷积神经网络和递归神经网络，已经成为当下流行的深度学习库。

本书讲述了如何在 Keras 库的帮助下，高效地解决在训练深度学习模型时遇到的各种问题。内容包括如何安装和设置 Keras，如何在 TensorFlow、Apache MXNet 和 CNTK 后端开发中使用 Keras 实现深度学习。

从加载数据到拟合和评估模型获得最佳性能，你将逐步解决在深度学习建模时可能遇到的所有问题。在本书的帮助下，你将实现卷积神经网络、递归神经网络、对抗网络等。除此之外，你还将学习如何训练这些模型以实现真实的图像处理和语言处理任务。

本书的最后，你将完成一个实例以进一步了解如何利用 Python 和 Keras 的强大功能实现有效的深度学习。

本书读者对象

本书适合数据科学家或机器学习专家，可以帮助他们解决在训练深度学习模型时遇到的常见问题。阅读本书前，需要对 Python 有基本的了解，并了解机器学习和神经网络的内容。

本书涵盖的内容

第 1 章介绍了 Keras 的安装和设置过程以及如何配置 Keras。

第 2 章介绍了使用 CIFAR-10、CIFAR-100 或 MNIST 等数据集，以及用于图像分类的其他数据集和模型。

第 3 章介绍了使用 Keras 的各种预处理和优化技术，优化技术包括 TFOptimizer、AdaDelta 等。

第 4 章详细描述了不同的 Keras 层，包括递归层和卷积层等。

第 5 章通过宫颈癌分类和数字识别数据集的实例，详细解释如何使用卷积神经网络算法。

第 6 章包括基本的生成式对抗网络（GAN）和边界搜索 GAN。

第 7 章涵盖了递归神经网络的基础，以便实现基于历史数据集的 Keras。

第 8 章包括使用 Keras 进行单词分析和情感分析的 NLP 基础知识。

第 9 章展示了如何在 Amazon 评论数据集中使用 Keras 模型进行文本概述。

第 10 章侧重于使用 Keras 设计和开发强化学习模型。

阅读本书须知

读者应该掌握 Keras 和深度学习的基本知识。

排版约定

本书中包含许多排版约定。

文本中的代码元素、数据库表名、文件夹名、文件名、文件扩展名、路径名、用户输入和 Twitter 句柄都采用代码字体表示。举个例子："最后，我们将所有评论保存到 `pickle` 文件中。"

代码段如下所示：

```
stories = list()
for i, text in enumerate(clean_texts):
    stories.append({'story': text, 'highlights': clean_summaries[i]})
```

当我们希望引起你对代码块的特定部分的注意时，相关的行或代码会以粗体显示：

```
from keras.datasets import cifar10
```

命令行的输入、输出表示如下：

```
sudo apt-get install graphviz
```

粗体：表示新术语、重要单词或词组。

> 警告或重要说明。

> 提示或小技巧。

示例代码及彩图下载

本书的示例代码及所有截图和样图，可以从 http://www.packtpub.com 通过个人账号下载，也可以访问华章图书官网 http://www.hzbook.com，通过注册并登录个人账号下载。

目 录

译者序
审校者简介
前言

第1章 Keras安装 ·················· 1
1.1 引言 ························· 1
1.2 在Ubuntu 16.04上安装Keras ······ 1
    1.2.1 准备工作 ················ 2
    1.2.2 怎么做 ·················· 2
1.3 在Docker镜像中使用Jupyter Notebook安装Keras ············ 7
    1.3.1 准备工作 ················ 7
    1.3.2 怎么做 ·················· 7
1.4 在已激活GPU的Ubuntu 16.04上安装Keras ················ 9
    1.4.1 准备工作 ················ 9
    1.4.2 怎么做 ················· 10

第2章 Keras数据集和模型 ········ 13
2.1 引言 ························ 13
2.2 CIFAR-10数据集 ·············· 13
2.3 CIFAR-100数据集 ············· 15
2.4 MNIST数据集 ················ 17
2.5 从CSV文件加载数据 ·········· 18
2.6 Keras模型入门 ··············· 19
    2.6.1 模型的剖析 ············· 19
    2.6.2 模型类型 ··············· 19
2.7 序贯模型 ···················· 20
2.8 共享层模型 ·················· 27
    2.8.1 共享输入层简介 ········· 27
    2.8.2 怎么做 ················· 27
2.9 Keras函数API ················ 29
    2.9.1 怎么做 ················· 29
    2.9.2 示例的输出 ············· 31
2.10 Keras函数API——链接层 ···· 31
2.11 使用Keras函数API进行图像分类 ······················ 32

第3章 数据预处理、优化和可视化 ···················· 36
3.1 图像数据特征标准化 ·········· 36
    3.1.1 准备工作 ··············· 36
    3.1.2 怎么做 ················· 37
3.2 序列填充 ···················· 39

- 3.2.1 准备工作 ……………… 39
- 3.2.2 怎么做 ………………… 39
- 3.3 模型可视化 ………………… 41
  - 3.3.1 准备工作 ……………… 41
  - 3.3.2 怎么做 ………………… 41
- 3.4 优化 ………………………… 43
- 3.5 示例通用代码 ……………… 43
- 3.6 随机梯度下降优化法 ……… 44
  - 3.6.1 准备工作 ……………… 44
  - 3.6.2 怎么做 ………………… 44
- 3.7 Adam 优化算法 …………… 47
  - 3.7.1 准备工作 ……………… 47
  - 3.7.2 怎么做 ………………… 47
- 3.8 AdaDelta 优化算法 ………… 50
  - 3.8.1 准备工作 ……………… 51
  - 3.8.2 怎么做 ………………… 51
- 3.9 使用 RMSProp 进行优化 …… 54
  - 3.9.1 准备工作 ……………… 54
  - 3.9.2 怎么做 ………………… 54

第 4 章 使用不同的 Keras 层实现分类 …………………………… 58
- 4.1 引言 ………………………… 58
- 4.2 乳腺癌分类 ………………… 58
- 4.3 垃圾信息检测分类 ………… 66

第 5 章 卷积神经网络的实现 …… 73
- 5.1 引言 ………………………… 73
- 5.2 宫颈癌分类 ………………… 73
  - 5.2.1 准备工作 ……………… 74
  - 5.2.2 怎么做 ………………… 74
- 5.3 数字识别 …………………… 84
  - 5.3.1 准备工作 ……………… 84
  - 5.3.2 怎么做 ………………… 85

第 6 章 生成式对抗网络 ………… 89
- 6.1 引言 ………………………… 89
- 6.2 基本的生成式对抗网络 …… 90
  - 6.2.1 准备工作 ……………… 91
  - 6.2.2 怎么做 ………………… 91
- 6.3 边界搜索生成式对抗网络 … 98
  - 6.3.1 准备工作 ……………… 99
  - 6.3.2 怎么做 ………………… 100
- 6.4 深度卷积生成式对抗网络 … 106
  - 6.4.1 准备工作 ……………… 107
  - 6.4.2 怎么做 ………………… 108

第 7 章 递归神经网络 …………… 116
- 7.1 引言 ………………………… 116
- 7.2 用于时间序列数据的简单 RNN ………………………… 117
  - 7.2.1 准备工作 ……………… 118
  - 7.2.2 怎么做 ………………… 119
- 7.3 时间序列数据的 LSTM 网络 … 128
  - 7.3.1 LSTM 网络 …………… 128
  - 7.3.2 LSTM 记忆示例 ……… 129
  - 7.3.3 准备工作 ……………… 129
  - 7.3.4 怎么做 ………………… 129

7.4 使用 LSTM 进行时间序列预测 ·············· 133
　7.4.1 准备工作 ·············· 134
　7.4.2 怎么做 ·············· 135
7.5 基于 LSTM 的等长输出序列到序列学习 ·············· 143
　7.5.1 准备工作 ·············· 143
　7.5.2 怎么做 ·············· 144

第 8 章 使用 Keras 模型进行自然语言处理 ·············· 150
8.1 引言 ·············· 150
8.2 词嵌入 ·············· 150
　8.2.1 准备工作 ·············· 151
　8.2.2 怎么做 ·············· 151
8.3 情感分析 ·············· 157
　8.3.1 准备工作 ·············· 157
　8.3.2 怎么做 ·············· 159

　8.3.3 完整代码清单 ·············· 162

第 9 章 基于 Keras 模型的文本摘要 ·············· 164
9.1 引言 ·············· 164
9.2 评论的文本摘要 ·············· 164
　9.2.1 怎么做 ·············· 165
　9.2.2 参考资料 ·············· 172

第 10 章 强化学习 ·············· 173
10.1 引言 ·············· 173
10.2 使用 Keras 进行《CartPole》游戏 ·············· 174
10.3 使用竞争 DQN 算法进行《CartPole》游戏 ·············· 181
　10.3.1 准备工作 ·············· 183
　10.3.2 怎么做 ·············· 187

CHAPTER 1

第 1 章

Keras 安装

本章包括以下内容：
- 在 Ubuntu 16.04 上安装 Keras
- 在 Docker 镜像中使用 Jupyter Notebook 安装 Keras
- 在已激活 GPU 的 Ubuntu 16.04 上安装 Keras

1.1 引言

在本章中，我们将讨论如何在 Ubuntu 和 CentOS 上安装 Keras。本书使用的是 64 位的 Ubuntu 16.04（即 Canonical 公司于 2017 年 10 月 26 日发布的 Ubuntu 16.04 LTS 和 amd64 xenial 镜像）。

1.2 在 Ubuntu 16.04 上安装 Keras

在安装 Keras 之前，我们必须安装 Theano 和 TensorFlow 软件包及其依赖项。除此之外，请确认你的操作系统已经安装了 Python。接下来是 Python 的安装过程介绍。

> Conda 是一个运行在多个 OS（Windows、macOS 和 Linux）上的开源软件包管理系统和环境管理系统。用于在 Python 环境下安装多个版本的软件包及其依赖项，并能在本地计算机上对各个版本进行创建、保存、加载和环境切换。

1.2.1 准备工作

首先,你需要确保在本地或云端具有完整的 Ubuntu 16.04 操作系统。

1.2.2 怎么做

接下来将介绍在安装 Keras 之前必须安装的各个组件。

安装 miniconda

首先,为了更方便地安装所需软件包,你需要先进行 miniconda 的安装。miniconda 是 conda 软件包管理器的精简版本,可以用它进行 Python 虚拟环境的创建。

> 💡 建议读者安装 Python 2.7 或 Python 3.4,Python = 2.7 * 或 (> = 3.4 和 <3.6),并安装 Python 开发包(Linux 发行的 `python-dev` 或 `python-devel`)。本书将基于 Python 2.7 进行讲解。

1. 为安装 miniconda,首先从 continuum 版本库下载所需 sh 安装文件:

```
wget
https://repo.continuum.io/miniconda/Miniconda2-latest-Linux-x86_64.
sh
chmod 755 Miniconda2-latest-Linux-x86_64.sh
./Miniconda2-latest-Linux-x86_64.sh
```

2. 成功安装 conda 后,我们就可以用它来安装 Theano、TensorFlow 和 Keras 的依赖项。

安装 numpy 和 scipy

`numpy` 和 `scipy` 是进行 Theano 安装的前提条件,建议安装以下版本:

- NumPy:不低于 1.9.1 且不高于 1.12。
- SciPy:不低于 0.14 且低于 0.17.1,用于处理稀疏矩阵和 Theano 中支持的一些特殊函数时强烈推荐,否则 SciPy 0.8 以上版本即可以满足需求。
- 建议安装 BLAS(具有 Level 3 功能):可通过 `conda` 与 `mkl-service` 包免费获得 MKL 数据库,在该库内找到 BLAS。

> ℹ️ 基本线性代数子程序库(BLAS)提供规范的线性代数运算程序,程序采用 C 或 Fortran 编写。包括例如向量加法、标量乘法、点积、线性组合和矩阵乘法,其中 Level 3 对应矩阵与矩阵的乘法运算。

1. 执行以下命令安装 `numpy` 和 `scipy`（确保 `conda` 在你的 `PATH` 中）：

```
conda install numpy
conda install scipy
```

`scipy` 安装时的输出如下所示，注意 `libgfortran` 安装也是 `scipy` 安装中的一部分：

```
Fetching package metadata ..........
Solving package specifications: .
Package plan for installation in environment
/home/ubuntu/miniconda2:
```

2. 以下软件包也同时被安装：

```
libgfortran-ng: 7.2.0-h9f7466a_2
scipy:          1.0.0-py27hf5f0f52_0
Proceed ([y]/n)?
libgfortran-ng 100%
|################################################################|
Time: 0:00:00  36.60 MB/s
scipy-1.0.0-py 100%
|################################################################|
Time: 0:00:00  66.62 MB/s
```

安装 mkl

1. `mkl` 是用于 Intel 及兼容处理器的数学库。它是 `numpy` 的一部分，但我们需要在安装 Theano 和 TensorFlow 之前先安装它：

```
conda install mkl
```

安装时的输出如下所示，本例中，`miniconda2` 已经安装了最新版的 `mkl`：

```
Fetching package metadata ..........
Solving package specifications: .
# All requested packages already installed.
# packages in environment at /home/ubuntu/miniconda2:
#
mkl   2018.0.1  h19d6760_4
```

2. 当完成上述安装后，就可以开始安装 TensorFlow 了。

安装 TensorFlow

1. 通过执行以下命令，利用 `conda` 开始安装 `tensorflow`：

```
conda install -c conda-forge tensorflow
```

如下所示，执行该命令可获取元数据并安装一系列的程序包：

```
Fetching package metadata .............
Solving package specifications: .
Package plan for installation in environment
/home/ubuntu/miniconda2:
```

2. 继续安装以下程序包：

```
bleach:      1.5.0-py27_0         conda-forge
funcsigs:    1.0.2-py_2           conda-forge
futures:     3.2.0-py27_0         conda-forge
html5lib:    0.9999999-py27_0     conda-forge
markdown:    2.6.9-py27_0         conda-forge
mock:        2.0.0-py27_0         conda-forge
pbr:         3.1.1-py27_0         conda-forge
protobuf:    3.5.0-py27_0         conda-forge
tensorboard: 0.4.0rc3-py27_0      conda-forge
tensorflow:  1.4.0-py27_0         conda-forge
webencodings: 0.5-py27_0          conda-forge
werkzeug:    0.12.2-py_1          conda-forge
```

3. 高版本的程序包将取代低版本的包：

```
conda:     4.3.30-py27h6ae6dc7_0 --> 4.3.29-py27_0 conda-forge
conda-env: 2.6.0-h36134e3_1      --> 2.6.0-0 conda-forge
Proceed ([y]/n)? y
conda-env-2.6. 100%
|###############################################################|
Time: 0:00:00   1.67 MB/s
...
mock-2.0.0-py2 100%
|###############################################################|
Time: 0:00:00  26.00 MB/s
conda-4.3.29-p 100%
|###############################################################|
Time: 0:00:00  27.46 MB/s
```

4. 使用以下命令创建名为 hello_tf.py 的新文件，来测试是否成功安装了 TensorFlow：

```
vi hello_tf.py
```

5. 将以下代码添加到此文件并保存：

```python
import tensorflow as tf
hello = tf.constant('Greetings, TensorFlow!')
sess = tf.Session()
print(sess.run(hello))
```

6. 在命令行执行已创建的文件：

`python hello_tf.py`

若得到如下输出则证明你已经成功安装了 TensorFlow 库：

`Greetings, TensorFlow!`

安装 Keras

> `conda-forge` 是 GitHub 平台上带有 `conda` 的工具。

1. 使用 conda-forge 上的 cohda 安装 Keras。
2. 在终端上执行以下命令：

`conda install -c conda-forge keras`

以下输出表明 Keras 已成功安装：

```
Fetching package metadata .............
Solving package specifications: .
Package plan for installation in environment
/home/ubuntu/miniconda2:
```

同时，以下软件包会被自动安装：

```
h5py: 2.7.1-py27_2 conda-forge
hdf5: 1.10.1-1 conda-forge
keras: 2.0.9-py27_0 conda-forge
libgfortran: 3.0.0-1
pyyaml: 3.12-py27_1 conda-forge
Proceed ([y]/n)? y
libgfortran-3. 100%
|################################################################|
Time: 0:00:00 35.16 MB/s
hdf5-1.10.1-1. 100%
|################################################################|
Time: 0:00:00 34.26 MB/s
pyyaml-3.12-py 100%
|################################################################|
Time: 0:00:00 60.08 MB/s
h5py-2.7.1-py2 100%
|################################################################|
Time: 0:00:00 58.54 MB/s
keras-2.0.9-py 100%
|################################################################|
Time: 0:00:00 45.92 MB/s
```

3. 使用以下代码检验 Keras 的安装：

```
$ python
Python 2.7.14 |Anaconda, Inc.| (default, Oct 16 2017, 17:29:19)
```

4. 执行以下命令以验证 Keras 是否成功安装：

```
> from keras.models import Sequential
Using TensorFlow backend.
>>>
```

可以看到，此时 Keras 基于 TensorFlow 的后端。

在 Theano 后端上运行 Keras

1. 通过修改默认配置（`keras.json` 文件）将 Keras 的后端由 TensorFlow 更改为 Theano：

```
vi .keras/keras.json
```

默认文件的内容如下：

```
{ "image_data_format": "channels_last",
  "epsilon": 1e-07,
  "floatx": "float32",
  "backend": "tensorflow"
}
```

2. 修改后的文件内容如下，其中"backend"已更改为"theano"：

```
{ "image_data_format": "channels_last",
  "epsilon": 1e-07,
  "floatx": "float32",
  "backend": "theano"
}
```

3. 运行 Python 控制台，并基于 Theano 后端从 `keras.model` 导入 `Sequential`：

```
$ python
Python 2.7.14 |Anaconda, Inc.| (default, Oct 16 2017, 17:29:19)
[GCC 7.2.0] on linux2
Type "help", "copyright", "credits" or "license" for more information.
>>> from keras.models import Sequential
```

可以看到，此时 Keras 的后端已改为 Theano。

至此，我们安装了 miniconda、Theano 和 TensorFlow 的所有依赖项，以及 TensorFlow

和 Theano 本身，而且成功安装了 Keras。最后，我们还介绍了如何将 Keras 的后端从 TensorFlow 更改为 Theano。

1.3 在 Docker 镜像中使用 Jupyter Notebook 安装 Keras

本节介绍如何安装 Docker 容器，在容器中运行 Keras，并使用 Jupyter 访问它。

1.3.1 准备工作

从 https://docs.docker.com/engine/installation/ 安装最新版本的 Docker CLI。

1.3.2 怎么做

在这一部分，我们将介绍如何安装 Docker 容器。

安装 Docker 容器

1. 在终端执行以下命令以运行 Docker 容器，用 `rajdeepd/jupyter-keras` 得到容器镜像：

```
docker run -d -p 8888:8888 rajdeepd/jupyter-keras start-notebook.sh
--NotebookApp.token=''
```

2. 此时已经在本地成功安装并激活了 Notebook，可通过执行 `docker ps -a` 查看其输出结果，如下所示：

```
CONTAINER ID        IMAGE                       COMMAND
CREATED             STATUS                      PORTS
NAMES
45998a5eea89        rajdeepd/jupyter-keras       "tini -- start-
not..."             About an hour ago Up About an hour
0.0.0.0:8888->8888/tcp    admiring_wing
```

值得注意的是，此时主机端口 8888 已映射到容器端口 8888。

3. 通过 URL `http://localhost:8888` 打开浏览器：

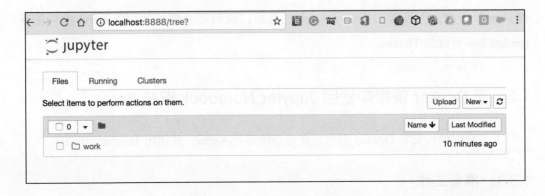

Jupyter处于运行状态，此时你可以创建一个新的Notebook并运行具体的Keras代码。

将本地卷映射到Docker容器

本节介绍如何将本地卷`$(pwd)/keras-samples`映射到容器工作目录中。

1. 执行`note -v flag`命令，该命令进行卷映射：

```
docker run -d -v /$(pwd)/keras-samples:/home/jovyan/work \
 -p 8888:8888 rajdeepd/jupyter-keras start-notebook.sh --
NotebookApp.token=''
```

如果此时进入URL，你会看到正在显示示例页面。

2. 进入`/$(pwd)/keras-samples`，你会发现host目录中有Notebooks，而且可利用Jupyter进行加载：

```
rdua1-ltm:keras-samples rdua$ pwd
/Users/rdua/personal/keras-samples
rdua1-ltm:keras-samples rdua$ ls
MNIST CNN.ipynb sample_one.ipynb
```

你可以尝试打开 MNIST CNN.ipynb，它是一个 Keras CNN 样本，我们将在后续章节中介绍更多相关信息。

在本节中，我们使用 Docker 镜像 rajdeepd/jupyter-keras 创建了一个 Keras 环境，并通过在主机环境中运行的 Jupyter 对其进行访问。

1.4 在已激活 GPU 的 Ubuntu 16.04 上安装 Keras

本节在已激活 NVIDIA GPU 的 Ubuntu 16.04 上进行 Keras 的安装。

1.4.1 准备工作

以支持 GPU 的 AWS EC2 实例为例，准备安装支持 GPU 的 TensorFlow 和 Keras。启动以下亚马逊机器镜像（AMI）：Ubuntu Server 16.04 LTS（HVM），SSD Volume Type - ami-aa2ea6d0：

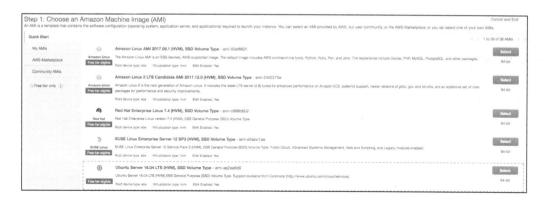

这是支持 Ubuntu 16.04，64 位的 AMI，它具有 SSD 卷类型，选择适当的实例类型：g3.4xlarge。

启动 VM 后，将用于 SSH 验证的相应密钥分配给它，本例使用的是一个预存密钥：

通过 SSH 连接实例：

```
ssh -i aws/rd_app.pem ubuntu@34.201.110.131
```

1.4.2 怎么做

1. 运行以下命令以 update 和 upgrade 操作系统：

```
sudo apt-get update
sudo apt-get upgrade
```

2. 安装 gcc 编译器并配置工具：

```
sudo apt install gcc
sudo apt install make
```

安装 cuda

1. 执行以下命令安装 cuda：

```
sudo apt-get install -y cuda
```

2. 运行以下基本程序检查 cuda 是否成功安装：

```
ls /usr/local/cuda-8.0
bin extras lib64 libnvvp nvml README share targets version.txt
doc include libnsight LICENSE nvvm samples src tools
```

3. 完成本地编译后运行一个 cuda 样例：

```
export PATH=/usr/local/cuda-8.0/bin${PATH:+:${PATH}}
export LD_LIBRARY_PATH=/usr/local/cuda-8.0/lib64\${LD_LIBRARY_PATH:+:${LD_LIBRARY_PATH}}
cd /usr/local/cuda-8.0/samples/5_Simulations/nbody
```

4. 编译并运行样例：

```
sudo make

./nbody
```

你将看到类似下述列表的输出：

```
Run "nbody -benchmark [-numbodies=<numBodies>]" to measure
performance.
 -fullscreen (run n-body simulation in fullscreen mode)
 -fp64 (use double precision floating point values for simulation)
 -hostmem (stores simulation data in host memory)
 -benchmark (run benchmark to measure performance)
 -numbodies=<N> (number of bodies (>= 1) to run in simulation)
 -device=<d> (where d=0,1,2.... for the CUDA device to use)
 -numdevices=<i> (where i=(number of CUDA devices > 0) to use for
simulation)
 -compare (compares simulation results running once on the default
GPU and once on the CPU)
 -cpu (run n-body simulation on the CPU)
 -tipsy=<file.bin> (load a tipsy model file for simulation)
```

5. 接下来安装 cudnn，这是一个来自 NVIDIA 的深度学习库，访问 https://developer.nvidia.com/cudnn 以找到更多相关信息。

安装 cudnn

1. 从 NVIDIA 网站（https://developer.nvidia.com/rdp/assets/cudnn-8.0-linux-x64-v5.0-ga-tgz）下载 cudnn 并解压缩二进制文件：

```
tar xvf cudnn-8.0-linux-x64-v5.1.tgz
```

> 请注意，执行这一步需要有一个 NVIDIA 开发者账户。

将下载的 `.tgz` 文件解压缩后，会获得以下输出：

```
cuda/include/cudnn.h
cuda/lib64/libcudnn.so
cuda/lib64/libcudnn.so.5
cuda/lib64/libcudnn.so.5.1.10
cuda/lib64/libcudnn_static.a
```

2. 将这些文件复制到 `/usr/local` 文件夹，如下所示：

```
sudo cp cuda/include/cudnn.h /usr/local/cuda/include
sudo cp cuda/lib64/libcudnn* /usr/local/cuda/lib64

sudo chmod a+r /usr/local/cuda/include/cudnn.h
/usr/local/cuda/lib64/libcudnn*
```

安装 NVIDIA CUDA 分析工具与界面开发文件

使用以下代码安装 TensorFlow GPU 安装所需的 NVIDIA CUDA 分析工具与界面开发文件：

```
sudo apt-get install libcupti-dev
```

安装 GPU 版本的 TensorFlow

执行以下命令以安装 GPU 版本的 TensorFlow：

```
sudo pip install tensorflow-gpu
```

安装 Keras

对于 Keras，使用示例命令，与用于 GPU 的安装命令类似：

```
sudo pip install keras
```

本小节中，我们学习了如何在配置了 CUDA 和 cuDNN 的 TensorFlow（GPU 版本）的基础上安装 Keras。

CHAPTER 2

第 2 章

Keras 数据集和模型

本章包括以下内容：
- CIFAR-10 数据集
- CIFAR-100 数据集
- MNIST 数据集
- 从 CSV 文件加载数据
- Keras 模型入门
- 序贯模型
- 共享层模型
- Keras 函数 API
- Keras 函数 API——链接层
- 使用 Keras 函数 API 进行图像分类

2.1 引言

在本章中，我们将介绍 Keras 中的各种默认数据集以及如何加载和使用它们。

2.2 CIFAR-10 数据集

从 https://www.cs.toronto.edu/~kriz/cifar-10-python.tar.gz 加载 CIFAR-10 小图像分类

数据集。CIFAR-10 数据集共有 60 000 张彩色图像，这些图像的分辨率为 32×32，分为 10 类，每类 6 000 张图。这里面有 50 000 张用于训练，构成了 5 个训练批次，每一批 10 000 张图。另外 10 000 张用于测试，单独构成一批，测试批次包含来自每个类的 1 000 张随机选择的图像。注意一个训练批次中的各类图像数量并不一定相同，总的训练样本包含来自每一类的 5 000 张图。如下图所示：

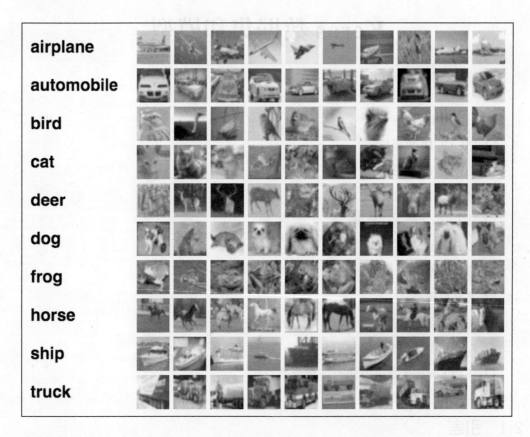

参考：https://www.cs.toronto.edu/~kriz/cifar.html。

怎么做

使用 Keras API 加载此数据集，并打印数据集的形状和大小：

```
from keras.datasets import cifar10
(X_train, y_train), (X_test, y_test) = cifar10.load_data()
```

```
print("X_train shape: " + str(X_train.shape))
print(y_train.shape)
print(X_test.shape)
print(y_test.shape)
```

首先，从前面的网站中下载文件：

```
Downloading data from
https://www.cs.toronto.edu/~kriz/cifar-10-python.tar.gz
8192/170498071 [..............................] - ETA: 22:43
40960/170498071 [..............................] - ETA: 9:12
106496/170498071 [..............................] - ETA: 5:27
237568/170498071 [..............................] - ETA: 3:11
286720/170498071 [..............................] - ETA: 4:39
...
170418176/170498071 [============================>.] - ETA: 0s
170467328/170498071 [============================>.] - ETA: 0s
170500096/170498071 [=============================] - 308s 2us/step
```

输出显示 `X_train` 包含 50 000 个大小为 32×32 的三通道图像，`y_train` 有 50 000 行和包含图像标签的一列。`X_test` 和 `y_test` 也有 10 000 个类似形状的数据：

```
X_train shape: (50000, 32, 32, 3)
y_train shape: (50000, 1)
X_test shape: (10000, 32, 32, 3)
y_test shape: (10000, 1)
```

在第 2.3 节，我们将了解如何加载 CIFAR-100 数据集。

2.3 CIFAR-100 数据集

训练数据集包含标记为 100 个类别的 50 000 个 32×32 像素彩色图像，以及 10 000 个测试图像。此数据集类似于 CIFAR-10，但它有 100 个类，每个类有 600 个图像（包括 500 个训练图像和 100 个测试图像）。CIFAR-100 中的 100 个类被分为 20 个超类。每个图像都带有一个粗标签（它所属的超类）和一个精细标签（它所属的类）。

CIFAR-100 中的类列表如下：

Superclass	Classes
aquatic mammals	beaver, dolphin, otter, seal, and whale
fish	aquarium fish, flatfish, ray, shark, and trout
flowers	orchids, poppies, roses, sunflowers, and tulips
food containers	bottles, bowls, cans, cups, and plates

(续)

Superclass	Classes
fruit and vegetables	apples, mushrooms, oranges, pears, and sweet peppers
household electrical devices	clock, computer keyboard, lamp, telephone, and television
household furniture	bed, chair, couch, table, and wardrobe
insects	bee, beetle, butterfly, caterpillar, and cockroach
large carnivores	bear, leopard, lion, tiger, and wolf
large man-made outdoor things	bridge, castle, house, road, and skyscraper
large natural outdoor scenes	cloud, forest, mountain, plain, and sea
large omnivores and herbivores	camel, cattle, chimpanzee, elephant, and kangaroo
medium-sized mammals	fox, porcupine, possum, raccoon, and skunk
non-insect invertebrates	crab, lobster, snail, spider, and worm
people	baby, boy, girl, man, and woman
reptiles	crocodile, dinosaur, lizard, snake, and turtle
small mammals	hamster, mouse, rabbit, shrew, and squirrel
trees	maple, oak, palm, pine, and willow
vehicles 1	bicycle, bus, motorcycle, pickup truck, and train
vehicles 2	lawn-mower, rocket, streetcar, tank, and tractor

参考：https://www.cs.toronto.edu/~kriz/cifar.html。

怎么做

让我们看看如何加载此数据集并打印其形状，CIFAR-100 数据集可通过 `keras.datasets.cifar100` 中的 `load_data()` 函数获得。

数据集从 https://www.cs.toronto.edu/~kriz/cifar-100-python.tar.gz 下载，这隐藏在以下实现中：

```
from keras.datasets import cifar100
(X_train, y_train), (X_test, y_test) = cifar100.load_data()
print("X_train shape: " + str(X_train.shape))
print("y_train shape: " + str(y_train.shape))
print("X_test shape: " + str(X_test.shape))
print("y_test shape: " + str(y_test.shape))
```

训练数据和测试数据的形状输出如下所示：

```
X_train shape: (50000, 32, 32, 3)
y_train shape: (50000, 1)
X_test shape: (10000, 32, 32, 3)
y_test shape: (10000, 1)
```

指定标签模式

可以通过 `load_data()` 函数进行标签模式的指定：

```
(X_train, y_train), (X_test, y_test) =
cifar100.load_data(label_mode='fine')
```

2.4 MNIST 数据集

MNIST 是一个包含 60 000 个 0～9 这十个数字的 28×28 像素灰度图像的数据集。MNIST 也包括 10 000 个测试集图像。数据集包含以下四个文件：

- `train-images-idx3-ubyte.gz`：训练集图像（9 912 422 字节），见 http://yann.lecun.com/exdb/mnist/train-images-idx3-ubyte.gz
- `train-labels-idx1-ubyte.gz`：训练集标签（28 881 字节），见 http://yann.lecun.com/exdb/mnist/train-labels-idx1-ubyte.gz
- `t10k-images-idx3-ubyte.gz`：测试集图像（16 48 877 字节），见 http://yann.lecun.com/exdb/mnist/t10k-images-idx3-ubyte.gz
- `t10k-labels-idx1-ubyte.gz`：测试集标签（4 542 字节），见 http://yann.lecun.com/exdb/mnist/t10k-labels-idx1-ubyte.gz

这些文件中的数据以 IDX 格式存储。IDX 文件格式是用于存储向量与多维度矩阵的文件格式，你可以在 http://www.fon.hum.uva.nl/praat/manual/IDX_file_format.html 上找到 IDX 格式的更多信息。

上图显示了 MNIST 数据集表示的图像。

怎么做

使用 `keras.datasets.mnist` 将 MNIST 数据加载到 numpy 数组中:

```
from keras.datasets import mnist
(X_train, y_train), (X_test, y_test) = mnist.load_data()
print("X_train shape: " + str(X_train.shape))
print("y_train shape: " + str(y_train.shape))
print("X_test shape: " + str(X_test.shape))
print("y_test shape: " + str(y_test.shape))
```

数据集的形状输出如下:

```
X_train shape: (60000, 28, 28)
y_train shape: (60000,)
X_test shape: (10000, 28, 28)
y_test shape: (10000,)
```

接下来介绍如何从 `.csv` 文件加载数据。

2.5 从 CSV 文件加载数据

除了预先存在的数据集之外,Keras 还可以直接从 numpy 数组中获取数据。

怎么做

可以从互联网上获取现成的 `.csv` 文件并使用它来加载 Keras 数据集:

```
dataset =
numpy.loadtxt("https://raw.githubusercontent.com/jbrownlee/Datasets/master/pima-indians-diabetes.data.csv", delimiter=",")
# split into input (X) and output (Y) variables
X = dataset[:,0:8]
Y = dataset[:,8]
```

请注意,数据集可以直接从 `.csv` 文件的 URL 加载。

代码输出如下:

```
[ 6. 148. 72. 35. 0. 33.6 0.627 50. ]
1.0
```

2.6 Keras 模型入门

本节介绍如何在 Keras 中创建一个基本模型。

2.6.1 模型的剖析

模型（Model）是 `Network` 的子类，它将训练和评估这样的例行程序添加到 `Network` 中。下图显示了各个类之间的关系。

> `Network` 不是开发人员直接使用的类，因此本节中的某些信息仅供你参考。

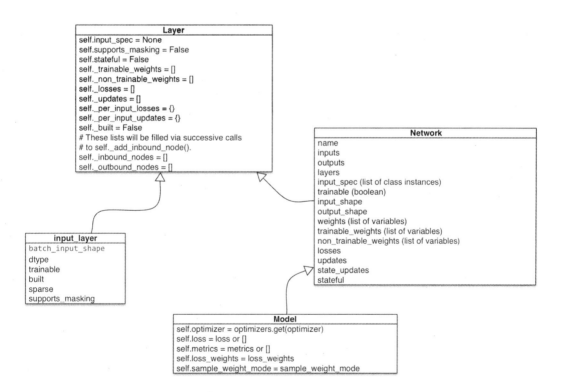

2.6.2 模型类型

Keras 有两种模型类型：

❑ 序贯模型

❏ 使用函数 API 创建的模型

2.7 序贯模型

可以通过将多个层堆叠并传递给 `Sequential` 的构造函数来创建序贯模型。

怎么做

创建基本序贯模型涉及指定一个或多个层。

创建序贯模型

我们将创建一个包含四层的序贯网络。

1. 第 1 层是全连接层（稠密层），其 `input_shape` 为 (`*`, `784`)，`output_shape` 为 (`*`, `32`)。

> 稠密层是一个根据一定规则密集连接的神经网络层，实现操作：*output = activation* (*dot* (*input*, *kernel*) + *bias*)。其中激活函数（*activation*）是传递激活参数的函数，核（*kernel*）是由层创建的权重矩阵，偏差（*bias*）是层创建的偏差向量（仅当 `use_bias` 为 `True` 时才适用）。

2. 第 2 层是激活层，将 `tanh` 激活函数用于激活输入张量：

```
keras.layers.Activation(activation)
```

`Activation` 也可以作为参数应用于稠密层：

```
model.add(Dense(64, activation='tanh'))
```

3. 第 3 层是一个稠密层，输出为 (`*`, `10`)。

4. 第 4 层具有应用 `softmax` 函数的 `Activation`：

```
Activation('softmax')
```

> 在数学中，`softmax` 函数也称为**归一化指数函数**，是逻辑函数的泛化，它将一个含任意实数的 K 维向量 z 压缩到另一个 K 维实向量 $\sigma(z)$。其中每个元素都在

（0～1）范围内，所有元素加起来为 1。如以下公式所示：

$$\sigma(z)_j = \frac{e^{z_j}}{\sum_{k=1}^{K} e^{z_k}}$$

```
from keras.models import Sequential
from keras.layers import Dense, Activation
model = Sequential([
  Dense(32, input_shape=(784,)),
  Activation('tanh'),
  Dense(10),
  Activation('softmax'),
])
print(model.summary())
```

5. 创建的模型如下：

```
Layer (type) Output Shape Param #
=================================================================
dense_1 (Dense) (None, 32) 25120
_____
activation_1 (Activation) (None, 32) 0
_____
dense_2 (Dense) (None, 10) 330
_____
activation_2 (Activation) (None, 10) 0
=================================================================
Total params: 25,450
Trainable params: 25,450
Non-trainable params: 0
```

`Sequential` 是 `Model` 的子类，还有一些其他方法将会在后续章节进行介绍。

编译模型

编译模型：

```
compile(optimizer, loss=None, metrics=None, loss_weights=None,
sample_weight_mode=None, weighted_metrics=None, target_tensors=None)
```

有关每个参数含义的详细信息，请参阅文档 https://keras.io/models/sequential/#the-sequential-model-api。

训练模型

此方法用于以指定次数（数据集上的迭代）训练模型：

```
fit(x=None, y=None, batch_size=None, epochs=1, verbose=1, callbacks=None,
validation_split=0.0, validation_data=None, shuffle=True,
class_weight=None, sample_weight=None, initial_epoch=0,
steps_per_epoch=None, validation_steps=None)
```

有关每个参数含义的详细信息，请参阅文档 https://keras.io/models/sequential/#the-sequential-model-api。

评估模型

evaluate 方法用于评估模型指标，使用批处理完成：

```
evaluate(x=None, y=None, batch_size=None, verbose=1, sample_weight=None,
steps=None)
```

模型预测

调用以下 `predict` API 进行预测，它返回一个 `numpy` 数组：

```
predict(x, batch_size=None, verbose=0, steps=None)
```

接下来给出一个结合使用这些 API 的实例。

把它们结合

我们将使用皮马印第安人糖尿病数据集。

> 该数据集最初来自美国糖尿病、消化系统和肾脏疾病研究所，目标是使用数据集中包括的测量值来诊断性地预测患者是否患有糖尿病。不过这些数据集有一定的局限性，例如，数据集中的所有患者都是大于 21 岁的女性皮马印第安人。数据集由若干医学预测变量和一个目标变量（结果）组成，预测变量包括患者的怀孕次数、BMI、胰岛素水平、年龄等。

```
from keras.models import Sequential
from keras.layers import Dense
import numpy
# fix random seed for reproducibility
numpy.random.seed(7)
# load pima indians dataset
dataset = numpy.loadtxt("data/diabetes.csv", delimiter=",", skiprows=1)
# split into input (X) and output (Y) variables
X = dataset[:,0:8]
Y = dataset[:,8]
# create model
```

```python
model = Sequential()
model.add(Dense(12, input_dim=8, activation='relu'))
model.add(Dense(8, activation='relu'))
model.add(Dense(1, activation='sigmoid'))
# Compile model
model.compile(loss='binary_crossentropy', optimizer='adam',
metrics=['accuracy'])
# Fit the model
model.fit(X, Y, epochs=150, batch_size=10)
# evaluate the model
scores = model.evaluate(X, Y)
print("\n%s: %.2f%%" % (model.metrics_names[1], scores[1]*100))
```

数据集的形状为（768，9）。

数据集的值包括：

	0	1	2	3	4	5	6	7	8
0	6.00000	148.00000	72.00000	35.00000	0.00000	33.60000	0.62700	50.00000	1.00000
1	1.00000	85.00000	66.00000	29.00000	0.00000	26.60000	0.35100	31.00000	0.00000
2	8.00000	183.00000	64.00000	0.00000	0.00000	23.30000	0.67200	32.00000	1.00000
3	1.00000	89.00000	66.00000	23.00000	94.00000	28.10000	0.16700	21.00000	0.00000
4	0.00000	137.00000	40.00000	35.00000	168.00000	43.10000	2.28800	33.00000	1.00000
5	5.00000	116.00000	74.00000	0.00000	0.00000	25.60000	0.20100	30.00000	0.00000

X 的值，为第 0 到第 7 列：

	0	1	2	3	4	5	6	7
0	6.00000	148.00000	72.00000	35.00000	0.00000	33.60000	0.62700	50.00000
1	1.00000	85.00000	66.00000	29.00000	0.00000	26.60000	0.35100	31.00000
2	8.00000	183.00000	64.00000	0.00000	0.00000	23.30000	0.67200	32.00000
3	1.00000	89.00000	66.00000	23.00000	94.00000	28.10000	0.16700	21.00000
4	0.00000	137.00000	40.00000	35.00000	168.00000	43.10000	2.28800	33.00000
5	5.00000	116.00000	74.00000	0.00000	0.00000	25.60000	0.20100	30.00000

Y 的值是数据集的第 8 列，如下图所示：

	0
0	1.00000
1	0.00000
2	1.00000
3	0.00000
4	1.00000
5	0.00000

模型内部检验

通过在调试器中进行模型检验,可以在调用 `compile` 方法之前得到如下模型属性:

```
input=Tensor("dense_1_input:0", shape=(?, 8), dtype=float32)
input_names=<class 'list'>: ['dense_1_input']
input_shape=<class 'tuple'>: (None, 8)
inputs=<class 'list'>: [<tf.Tensor 'dense_1_input:0' shape=(?, 8)
dtype=float32>]
layers=<class 'list'>: [
<keras.layers.core.Dense object at 0x7fdbcbb444a8>,
<keras.layers.core.Dense object at 0x7fdbcbb05c50>,
<keras.layers.core.Dense object at 0x7fdbcbb05cf8>]
output=Tensor("dense_3/Sigmoid:0", shape=(?, 1), dtype=float32)
output_names=<class 'list'>: ['dense_3']
output_shape=<class 'tuple'>: (None, 1)
outputs-<class 'list'>: [<tf.Tensor 'dense_3/Sigmoid:0' shape=(?, 1)
dtype=float32>]
trainable_weights=<class 'list'>:
 [<tf.Variable 'dense_1/kernel:0' shape=(8, 12) dtype=float32_ref>,
  <tf.Variable 'dense_1/bias:0' shape=(12,) dtype=float32_ref>,
  <tf.Variable 'dense_2/kernel:0' shape=(12, 8) dtype=float32_ref>,
  <tf.Variable 'dense_2/bias:0' shape=(8,) dtype=float32_ref>,
  <tf.Variable 'dense_3/kernel:0' shape=(8, 1) dtype=float32_ref>,
  <tf.Variable 'dense_3/bias:0' shape=(1,) dtype=float32_ref>]
  weights=<class 'list'>:
[<tf.Variable 'dense_1/kernel:0' shape=(8, 12) dtype=float32_ref>,
 <tf.Variable 'dense_1/bias:0' shape=(12,) dtype=float32_ref>,
 <tf.Variable 'dense_2/kernel:0' shape=(12, 8) dtype=float32_ref>,
 <tf.Variable 'dense_2/bias:0' shape=(8,) dtype=float32_ref>,
 <tf.Variable 'dense_3/kernel:0' shape=(8, 1) dtype=float32_ref>,
 <tf.Variable 'dense_3/bias:0' shape=(1,) dtype=float32_ref>]
```

模型内部编译

调用 `model.compile()` 时后台会进行如下操作:

从后端获得优化器。

以下是支持的优化器列表:

```
all_classes = {
'sgd': SGD,
'rmsprop': RMSprop,
'adagrad': Adagrad,
'adadelta': Adadelta,
'adam': Adam,
'adamax': Adamax,
'nadam': Nadam,
'tfoptimizer': TFOptimizer,
}
```

初始化损失函数

所用的损失函数是二元交叉熵损失。

> 交叉熵损失（也称为**对数损失**）用于测量模型的性能（分类模型），其输出是介于 0 ～ 1 之间的概率值，随着预测概率偏离实际值的程度变化而变化：

$$-(y \log (p) + (1-y) \log (1-p))$$

```
self.loss = loss or []
```

初始化所有输出的内部变量：

```
self._feed_outputs = []
self._feed_output_names = []
self._feed_output_shapes = []
self._feed_loss_fns = []
```

设置模型目标：

```
self._feed_targets.append(target)
 self._feed_outputs.append(self.outputs[i])
 self._feed_output_names.append(name)
 self._feed_output_shapes.append(shape)
 self._feed_loss_fns.append(self.loss_functions[i])
```

设置样本权重：

在编译之前，将以下值分配给样本权重和 `sample_weight_modes`：

```
sample_weights = []
sample_weight_modes = []
```

运行代码后，它会初始化为以下值：

```
Tensor("dense_3_sample_weights:0", shape=(?,), dtype=float32)
```

设置指标：

接下来，我们设置指标名称和 `metrics_tensors`，用于存储实际指标：

```
self.metrics_names = ['loss']
self.metrics_tensors = []
```

计算总损失和指标：

计算损失并添加到 `self.metrics_tensors`：

```
output_loss = weighted_loss(y_true, y_pred,
 sample_weight, mask)
...
self.metrics_tensors.append(output_loss)
self.metrics_names.append(self.output_names[i] + '_loss')
```

接下来，计算嵌套指标和 **nested_weighted_metrics**：

```
nested_metrics = collect_metrics(metrics, self.output_names)
nested_weighted_metrics = collect_metrics(weighted_metrics,
self.output_names)
```

初始化测试、训练和预测函数：

全部惰性初始化：

```
self.train_function = None
self.test_function = None
self.predict_function = None
```

对可训练权重进行排序：

最后，我们初始化可训练的权重：

```
trainable_weights = self.trainable_weights
self._collected_trainable_weights = trainable_weights
```

模型训练

调用 model.fit 进行模型训练，执行以下步骤。

数据验证：

将 validation_data 传递给 Keras 模型时，它必须包含两个参数（x_val, y_val）或三个参数（x_val, y_val 和 val_sample_weights）。

模型输出

上述代码中模型指标的最终输出显示如下：

```
 10/768 [..............................] - ETA: 0s - loss: 0.5371 - acc: 0.7000
400/768 [===============>..............] - ETA: 0s - loss: 0.4888 - acc: 0.7625
768/768 [==============================] - 0s 131us/step - loss: 0.4727 - acc: 0.7656
Epoch 150/150
 10/768 [..............................] - ETA: 0s - loss: 0.3373 - acc: 0.9000
470/768 [=================>............] - ETA: 0s - loss: 0.4534 - acc:
```

```
0.7894
 768/768 [==============================] - 0s 122us/step - loss: 0.4783 - acc: 0.7799
 32/768 [>.............................] - ETA: 0s
 768/768 [==============================] - 0s 51us/step
 acc: 77.60%
```

2.8 共享层模型

Keras 中可以多层共享一个层的输出。例如输入中可以存在多个不同的特征提取层，或者可以使用多个层来预测特征提取层的输出。

下面进行示例介绍。

2.8.1 共享输入层简介

本节将介绍具有不同大小内核的多个卷积层如何解译同一图像的输入。该模型采用尺寸为 32×32×3 像素的彩色 CIFAR 图像。有两个共享此输入的 CNN 特征提取子模型，其中一个内核大小为 4，另一个内核大小为 8。这些特征提取子模型的输出被平展为向量、然后串联成为一个长向量，并在最终输出层进行二进制分类之前，将其传递到全连接层以进行解译。

以下为模型拓扑：
- 一个输入层
- 两个特征提取层
- 一个解译层
- 一个稠密输出层

2.8.2 怎么做

首先，我们需要使用 Keras API 定义适当的层，这里的关键 API 作用是创建合并层并使用它来创建解译层。

concatenate 函数

concatenate 函数用于合并两个模型，如以下代码所示：

```
# merge feature extractors
merge = concatenate([flat1, flat2])
# interpretation layer
hidden1 = Dense(512, activation='relu')(merge)
```

以下是完整的模型拓扑代码:

```
#input layer
visible = Input(shape=(32,32,3))
# first feature extractor
conv1 = Conv2D(32, kernel_size=4, activation='relu')(visible)
pool1 = MaxPooling2D(pool_size=(2, 2))(conv1)
flat1 = Flatten()(pool1)
# second feature extractor
conv2 = Conv2D(16, kernel_size=8, activation='relu')(visible)
pool2 = MaxPooling2D(pool_size=(2, 2))(conv2)
flat2 = Flatten()(pool2)
# merge feature extractors
merge = concatenate([flat1, flat2])
# interpretation layer
hidden1 = Dense(512, activation='relu')(merge)
# prediction output
output = Dense(10, activation='sigmoid')(hidden1)
model = Model(inputs=visible, outputs=output)
```

模型拓扑保存到文件中，并显示单个输入层如何馈送到两个特征提取层，如下图所示:

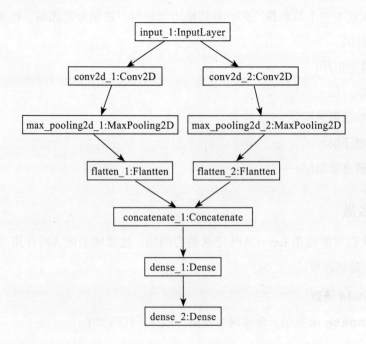

接下来，我们对这个模型进行编译，传入训练数据集并计算模型的准确度：

```
# Let's train the model using RMSprop
model.compile(loss='categorical_crossentropy',
 optimizer=opt,
 metrics=['accuracy'])
model.fit(x_train, y_train,
 batch_size=batch_size,
 epochs=epochs,
 validation_data=(x_test, y_test),
 shuffle=True)
scores = model.evaluate(x_test, y_test, verbose=1)
print('Test loss:', scores[0])
print('Test accuracy:', scores[1])
```

在第 2.9 节，我们将介绍如何使用 Keras 函数 API 创建模型并搭建更为复杂的拓扑。

2.9 Keras 函数 API

Keras 函数 API 通过函数创建每个层。

2.9.1 怎么做

1. 在使用函数 API 之前，你需要从 `keras` 包导入以下类：

```
from keras.layers.core import dense, Activation
```

2. 使用前面导入的层作为 `Sequential` 模型的一部分：

```
from keras.models import Sequential
from keras.layers.core import dense, Activation
model = Sequential([
  dense(32, input_dim=784),
  Activation("sigmoid"),
  dense(10),
  Activation("softmax"),
])
model.compile(loss="categorical_crossentropy", optimizer="adam")
```

3. 在 MNIST 上运行前面的基于函数 API 的模型：

```
from keras.utils import plot_model
from keras.layers import Flatten
from keras.models import Sequential
from keras.layers.core import Dense, Activation
from keras.datasets import mnist
```

```python
import keras

num_classes = 10
batch_size = 32
epochs = 10
batch_size = 128
num_classes = 10
epochs = 12

# input image dimensions
img_rows, img_cols = 28, 28

# the data, split between train and test sets
(x_train, y_train), (x_test, y_test) = mnist.load_data()

y_train = keras.utils.to_categorical(y_train, num_classes)
y_test = keras.utils.to_categorical(y_test, num_classes)


# input layer
model = Sequential([
  Flatten(input_shape=(28, 28)),
  Dense(32, input_dim=784),
  Activation("sigmoid"),
  Dense(10),
  Activation("softmax"),
])

# summarize layers
print(model.summary())

# plot graph
plot_model(model, to_file='shared_input_layer.png')

opt = keras.optimizers.rmsprop(lr=0.0001, decay=1e-6)

# Let's train the model using RMSprop
model.compile(loss='categorical_crossentropy',
              optimizer=opt,
              metrics=['accuracy'])


model.fit(x_train, y_train,
          batch_size=batch_size,
          epochs=epochs,
          validation_data=(x_test, y_test),
          shuffle=True)

scores = model.evaluate(x_test, y_test, verbose=1)
print('Test loss:', scores[0])
print('Test accuracy:', scores[1])
```

注意我们如何使用第一层在将输入馈入模型之前对其进行平展：

```
Flatten(input_shape=(28, 28)),
```

2.9.2 示例的输出

以下是示例输出（删减部分）：

```
50560/60000 [=========================>.....] - ETA: 0s - loss: 0.3895 -
acc: 0.8933
53504/60000 [============================>....] - ETA: 0s - loss: 0.3894 -
acc: 0.8929
57216/60000 [=============================>..] - ETA: 0s - loss: 0.3889 -
acc: 0.8928
60000/60000 [==============================] - 1s 17us/step - loss: 0.3886
- acc: 0.8928 - val_loss: 0.3846 - val_acc: 0.8925
   32/10000 [..............................] - ETA: 0s
 2592/10000 [======>.......................] - ETA: 0s
 5184/10000 [==============>...............] - ETA: 0s
 8064/10000 [=======================>......] - ETA: 0s
10000/10000 [==============================] - 0s 19us/step
Test loss: 0.3846480777263641
Test accuracy: 0.8925
```

2.10 Keras 函数 API——链接层

在函数模型中，我们必须创建一个输入层，用来指定输入数据的形状。输入层采用 shape 元素（元组）指定输入数据的维度。当输入数据是一维（例如对于多层感知器）时，必须为小批量训练数据留出空间，这是在训练网络时进行数据分割时就确定的，当输入是一维示例（如 32）时，shape 元组总是由最后一个开放维度来定义。

怎么做

创建第一层：

```
from keras.layers import Input
visible = Input(shape=(32,))
```

将各层连接在一起：

```
visible = Input(shape=(32,))
hidden = Dense(32)(visible)
```

Model 类

我们可以使用 `Model` 类来创建模型实例，如下代码片段所示：

```
from keras.models import Model
from keras.layers import Input
from keras.layers import Dense

visible = Input(shape=(32,))
hidden = Dense(32)(visible)
model = Model(inputs=visible, outputs=hidden)
```

定义具有多个输入和输出的模型：

```
model = Model(inputs=[a1, a2], outputs=[b1, b2, b3])
```

在上述代码中，我们定义了多层输入与多层输出，在第 2.11 节，我们将利用 Keras 函数 API 进行图像分类。

2.11 使用 Keras 函数 API 进行图像分类

我们已经学习了如何使用 Sequential 创建图像分类模型用于 MNIST，接下来将看看如何将卷积 API 与函数 API 一起使用。本节将重点介绍函数 API，卷积 API 的细节将会在本书的后面部分进行探讨。

怎么做

从批量的 MNIST 图像输入构建模型：

```
input_shape = (28, 28)
inputs = Input(input_shape)
print(input_shape + (1, ))
# add one more dimension for convolution
x = Reshape(input_shape + (1, ), input_shape=input_shape)(inputs)
conv1 = Conv2D(14, kernel_size=4, activation='relu')(x)
pool1 = MaxPooling2D(pool_size=(2, 2))(conv1)
conv2 = Conv2D(7, kernel_size=4, activation='relu')(pool1)
pool2 = MaxPooling2D(pool_size=(2, 2))(conv2)
flatten = Flatten()(pool2)
output = Dense(10, activation='sigmoid')(flatten)
model = Model(inputs=inputs, outputs=output)
```

首先 `input_shape` 为（28，28），用于定义输入层：

```
inputs = Input(input_shape)
```

然后我们为卷积添加另一个维度，并使用 **Reshape** 进行重新定义：

```
x = Reshape(input_shape + (1, ), input_shape=input_shape)(inputs)
```

定义两个卷积层和两个池化层：

```
conv1 = Conv2D(14, kernel_size=4, activation='relu')(x)
pool1 = MaxPooling2D(pool_size=(2, 2))(conv1)
conv2 = Conv2D(7, kernel_size=4, activation='relu')(pool1)
pool2 = MaxPooling2D(pool_size=(2, 2))(conv2)
```

这是模型创建：

```
model = Model(inputs=inputs, outputs=output)
# summarize layers
print(model.summary())
# plot graph
plot_model(model, to_file='convolutional_neural_network.png')
```

以下代码段为模型输出：

```
Using TensorFlow backend.
(28, 28, 1)
_____
Layer (type)                 Output Shape              Param #
=================================================================
input_1 (InputLayer)         (None, 28, 28)            0
_____
reshape_1 (Reshape)          (None, 28, 28, 1)         0
_____
conv2d_1 (Conv2D)            (None, 25, 25, 14)        238
_____
max_pooling2d_1 (MaxPooling2 (None, 12, 12, 14)        0
_____
conv2d_2 (Conv2D)            (None, 9, 9, 7)           1575
_____
max_pooling2d_2 (MaxPooling2 (None, 4, 4, 7)           0
_____
flatten_1 (Flatten)          (None, 112)               0
_____
dense_1 (Dense)              (None, 10)                1130
=================================================================
Total params: 2,943
Trainable params: 2,943
Non-trainable params: 0
```

图像分类模型示意图：

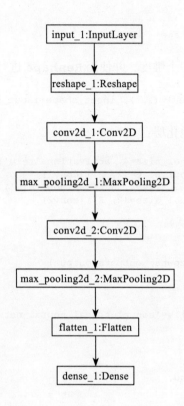

如下所示为模型的完整代码：

```
from keras.layers import Flatten
from keras.datasets import mnist
import keras
from keras.utils import plot_model
from keras.models import Model
from keras.layers import Input
from keras.layers import Dense
from keras.layers import Reshape
from keras.layers.convolutional import Conv2D
from keras.layers.pooling import MaxPooling2D

num_classes = 10
batch_size = 32
epochs = 10
batch_size = 128
num_classes = 10
epochs = 12

# input image dimensions
img_rows, img_cols = 28, 28

# the data, split between train and test sets
```

```python
(x_train, y_train), (x_test, y_test) = mnist.load_data()

y_train = keras.utils.to_categorical(y_train, num_classes)
y_test = keras.utils.to_categorical(y_test, num_classes)

input_shape = (28, 28)
inputs = Input(input_shape)
print(input_shape + (1, ))
# add one more dimension for convolution
x = Reshape(input_shape + (1, ), input_shape=input_shape)(inputs)
conv1 = Conv2D(14, kernel_size=4, activation='relu')(x)
pool1 = MaxPooling2D(pool_size=(2, 2))(conv1)
conv2 = Conv2D(7, kernel_size=4, activation='relu')(pool1)
pool2 = MaxPooling2D(pool_size=(2, 2))(conv2)
flatten = Flatten()(pool2)
output = Dense(10, activation='sigmoid')(flatten)
model = Model(inputs=inputs, outputs=output)
# summarize layers
print(model.summary())
# plot graph
plot_model(model, to_file='convolutional_neural_network.png')

opt = keras.optimizers.rmsprop(lr=0.0001, decay=1e-6)
# Let's train the model using RMSprop
model.compile(loss='categorical_crossentropy',
              optimizer=opt,
              metrics=['accuracy'])

model.fit(x_train, y_train, batch_size=batch_size, epochs=epochs,
validation_data=(x_test, y_test), shuffle=True)
scores = model.evaluate(x_test, y_test, verbose=1)
print('Test loss:', scores[0])
print('Test accuracy:', scores[1])
```

运行结果如下：

```
8480/10000 [=========================>.....] - ETA: 0s
8672/10000 [==========================>....] - ETA: 0s
8864/10000 [==========================>....] - ETA: 0s
9056/10000 [===========================>...] - ETA: 0s
9280/10000 [===========================>...] - ETA: 0s
9504/10000 [============================>..] - ETA: 0s
9760/10000 [=============================>.] - ETA: 0s
10000/10000 [==============================] - 2s 239us/step
Test loss: 3.944010387802124
Test accuracy: 0.5415
```

可以看出，它的准确率很低，针对这点将在第 5 章中进行调优。

CHAPTER 3

第 3 章

数据预处理、优化和可视化

本章将介绍以下内容：
- 图像数据特征标准化
- 序列填充
- 模型可视化
- 优化
- 示例通用代码
- 随机梯度下降优化法
- Adam 优化算法
- AdaDelta 优化算法
- RMSProp 优化算法

> 源代码链接：https://github.com/ml-resources/deeplearning-keras/tree/ed1/ch03。

3.1 图像数据特征标准化

本节将介绍如何使用 Keras 进行图像数据特征标准化。

3.1.1 准备工作

安装 Keras 与 Jupyter Notebook。

3.1.2 怎么做

使用 mnist 数据集。首先，绘制一个未经标准化的图像：

```
from keras.datasets import mnist
from matplotlib import pyplot

(X_train, y_train), (X_test, y_test) = mnist.load_data()
# create a grid of 3x3 images
for i in range(0, 9):
    ax = pyplot.subplot(330 + 1 + i)
    pyplot.tight_layout()
    ax.tick_params(axis='x', colors='white')
    ax.tick_params(axis='y', colors='white')
 pyplot.imshow(X_train[i], cmap=pyplot.get_cmap('gray'))
# show the plot
pyplot.show()
```

输出图像如下所示：

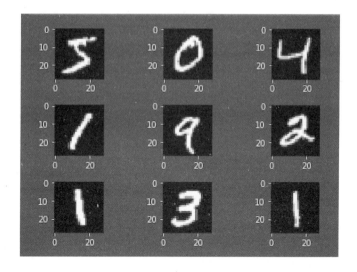

接下来使用 ImageDataGenerator 对该图进行特征标准化处理。

初始化 ImageDataGenerator

使用

```
keras.preprocessing.image.ImageDataGenerator(featurewise_center = False,
samplewise_center = False, featurewise_std_normalization = False, samplewise_std_
normalization = False)
```

初始化 ImageDataGenerator。

参数介绍如下：

- `featurewise_center`：布尔值，将数据集的输入均值设置为 0，按特征执行。
- `samplewise_center`：布尔值，将样本均值都初始化为 0。
- `featurewise_std_normalization`：布尔值，将输入除以数据集的标准差，按特征执行。
- `samplewise_std_normalization`：布尔值，将输入的每个样本除以其自身的标准差。

将这一初始化程序应用于 mnist 数据集：

```
from keras.preprocessing.image import ImageDataGenerator
from keras import backend as K
K.set_image_dim_ordering('th')

X_train = X_train.reshape(X_train.shape[0], 1, 28, 28)
X_test = X_test.reshape(X_test.shape[0], 1, 28, 28)
# convert from int to float
X_train = X_train.astype('float32')
X_test = X_test.astype('float32')
# define data preparation
datagen = ImageDataGenerator(featurewise_center=True,
featurewise_std_normalization=True,
 samplewise_center=True, samplewise_std_normalization=True)
# fit parameters from data
datagen.fit(X_train)
# configure batch size and retrieve one batch of images
for X_batch, y_batch in datagen.flow(X_train, y_train, batch_size=9):
    # create a grid of 3x3 images
    for i in range(0, 9):
        ax =pyplot.subplot(330 + 1 + i)
        pyplot.tight_layout()
        ax.tick_params(axis='x', colors='white')
        ax.tick_params(axis='y', colors='white')
        pyplot.imshow(X_batch[i].reshape(28, 28),
cmap=pyplot.get_cmap('gray'))
    # show the plot
    pyplot.show()
    break
```

输出为一个按特征及样本标准化并归一化处理后的 3×3 网格图，如下所示：

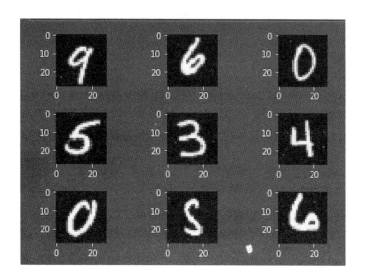

3.2 序列填充

本节将介绍如何利用 Keras 进行序列填充，序列填充常用于向 LSTM 网络分批发送序列时。

3.2.1 准备工作

导入函数：

```
from keras.preprocessing.sequence import pad_sequences
```

函数 `pad_sequences` 的定义如下：

```
pad_sequences(sequences, maxlen=None, dtype='int32', padding='pre',
truncating='pre', value=0.0)
```

3.2.2 怎么做

接下来介绍各种填充方式。

预填充，默认为 0.0 填充

首先，使用 `pad_sequences` 进行默认预填充：

```
from keras.preprocessing.sequence import pad_sequences
# define sequences
sequences = [
[1, 2, 3, 4],
[5, 6, 7],
[5]
]
# pad sequence
padded = pad_sequences(sequences)
print(padded)
```

`print` 语句输出的是填充到长度 4 的所有序列。

后填充

使用 `padding ='post'` 在较短的数组末尾填充 0.0，代码段如下所示：

```
padded_post = pad_sequences(sequences,padding='post')
print(padded_post)
```

截断填充

使用 `maxlen` 参数截断序列的第一个值或最后一个值：

```
padded_maxlen_truncating_pre = pad_sequences(sequences,maxlen=3,
truncating='pre')
print(padded_maxlen_truncating_pre)
```

输出如下所示：

```
[[2 3 4]
 [5 6 7]
 [0 0 5]]
```

可见第一行的第一个值被截去。

```
padded_maxlen_truncating_post = pad_sequences(sequences,maxlen=3,
truncating='post')
print(padded_maxlen_truncating_post)
```

输出如下所示：

```
[[1 2 3]
 [5 6 7]
 [0 0 5]]
```

可见第一行的最后一个值被截去。

非默认值填充

非默认值的填充（在本例中为 1.0）：

```
padded_value = pad_sequences(sequences, value=1.0)
print(padded_value)
```

如下是上述代码段的输出结果。其中，通过补 1 的方式使得每一行的长度都为 4：

```
[[1 2 3 4]
 [1 5 6 7]
 [1 1 1 5]]
```

3.3 模型可视化

对于较简单的模型，可利用简单的模型总结解决，但对于更复杂的拓扑结构，Keras 提供可视化模型的方法，即使用 `graphviz` 库。

3.3.1 准备工作

安装 `graphviz`：

```
sudo apt-get install graphviz
```

另外，安装 `pydot`，用于底层实现：

```
sudo pip install pydot
```

3.3.2 怎么做

接下来，创建一个简单的模型并调用 `plot_model`。

Keras 中通过 `plot_model()` 函数将神经网络绘制成图形。函数包括以下参数：

- `model`（必需）：要绘制的模型
- `to_file`（必需）：保存模型图的文件名称
- `show_shapes`（可选，默认为 `False`）：布尔值，用于显示每层的输出维度
- `show_layer_names`（可选，默认为 `True`）：布尔值，用于显示每层的名称

以下部分介绍如何使用 `plot_model`。

代码清单

创建一个包含两个 Dense 层的 Sequential 模型：

```
from keras.models import Sequential
from keras.layers import Dense
from keras.utils.vis_utils import plot_model
model = Sequential()
model.add(Dense(16, input_dim=1, activation='relu'))
model.add(Dense(16, activation='sigmoid'))
print(model.summary())
```

在 Keras 中可以使用 summary() 方法进行总结：

```
Using TensorFlow backend.
_____
Layer (type)              Output Shape         Param #
=================================================================
dense_1 (Dense)           (None, 16)           32
_____
dense_2 (Dense)           (None, 16)           272
=================================================================
Total params: 304
Trainable params: 304
Non-trainable params: 0
_____
```

对该模型调用 plot_model() 函数：

```
plot_model(model, to_file='model_plot.png', show_shapes=True,
show_layer_names=True)
```

输出图保存在以下文件中：

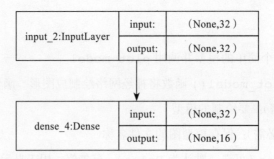

3.4 优化

通过优化使得 y 的预测值和实际值之间的损失函数值最小。Keras 支持各种优化技术，例如：

- SGD
- RMSProp
- Adam
- AdaDelta
- TFOptimizer
- AdaGrad

3.5 示例通用代码

以下代码可用于所有优化示例，我们导入与优化相关的类：

```
from __future__ import print_function
import keras
from keras.datasets import mnist
from keras.models import Sequential
from keras.layers import Dense, Dropout

batch_size = 128
num_classes = 10
epochs = 20

(x_train, y_train), (x_test, y_test) = mnist.load_data()

x_train = x_train.reshape(60000, 784)
x_test = x_test.reshape(10000, 784)
x_train = x_train.astype('float32')
x_test = x_test.astype('float32')
x_train /= 255
x_test /= 255
print(x_train.shape[0], 'train samples')
print(x_test.shape[0], 'test samples')
# convert class vectors to binary class matrices

y_train = keras.utils.to_categorical(y_train, num_classes)
y_test = keras.utils.to_categorical(y_test, num_classes)
```

3.6 随机梯度下降优化法

随机梯度下降（SGD），与批量梯度下降相反，它针对每个训练示例 $x^{(i)}$ 和输出 $y^{(i)}$ 进行参数更新：

$$\Theta = \Theta - \eta \nabla_\Theta j(\Theta, x^{(i)}, y^{(i)})$$

3.6.1 准备工作

在执行之前，需要在主代码段中添加示例通用代码。

3.6.2 怎么做

使用合适的网络拓扑创建一个序贯模型：

- 输入层：输入维度（*，784），输出维度（*，512）。
- 隐藏层：输入维度（*，512），输出维度（*，512）。
- 输出层：输入维度（*，512），输出维度（*，10）。

每层的激活函数：

- 第1层和第2层：`relu` 函数
- 第3层：`softmax` 函数

```
from keras.optimizers import SGD

y_train = keras.utils.to_categorical(y_train, num_classes)
y_test = keras.utils.to_categorical(y_test, num_classes)

model = Sequential()
model.add(Dense(512, activation='relu', input_shape=(784,)))
model.add(Dropout(0.2))
model.add(Dense(512, activation='relu'))
model.add(Dropout(0.2))
model.add(Dense(num_classes, activation='softmax'))

model.summary()
sgd = SGD(lr=0.01, decay=1e-6, momentum=0.9, nesterov=True)
model.compile(loss='categorical_crossentropy',
     optimizer=SGD(),metrics=['accuracy'])

#model.compile(loss='categorical_crossentropy',
#    optimizer=RMSprop(),
#    metrics=['accuracy'])
```

```
history = model.fit(x_train, y_train,
                    batch_size=batch_size,
                    epochs=epochs,
                    verbose=1,
                    validation_data=(x_test, y_test))
```

这里创建了一个具有两个隐藏层，丢弃率为 0.2 的网络。

上述代码的输出为：

```
Layer (type) Output Shape Param #
=================================================================
dense_10 (Dense) (None, 512) 401920
_____
dropout_7 (Dropout) (None, 512) 0
_____
dense_11 (Dense) (None, 512) 262656
_____
dropout_8 (Dropout) (None, 512) 0
_____
dense_12 (Dense) (None, 10) 5130
=================================================================
Total params: 669,706
Trainable params: 669,706
Non-trainable params: 0
```

计算模型的准确率和损失：

```
print(history.history.keys())
import matplotlib.pyplot as plt
%matplotlib inline
# summarize history for accuracy
plt.plot(history.history['acc'])
plt.plot(history.history['val_acc'])
plt.title('model accuracy')
plt.ylabel('accuracy')
plt.xlabel('epoch')
plt.legend(['train', 'test'], loc='upper left')
plt.show()
# summarize history for loss
plt.plot(history.history['loss'])
plt.plot(history.history['val_loss'])
plt.title('model loss')
plt.ylabel('loss')
plt.xlabel('epoch')
plt.legend(['train', 'test'], loc='upper left')
plt.show()
```

测试数据和训练数据的模型准确率曲线如下所示，均收敛到 95%：

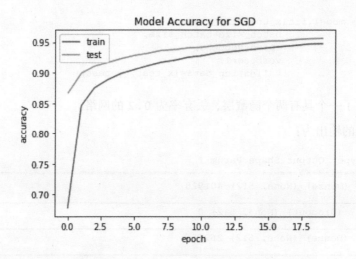

相应的模型损失如下图所示：

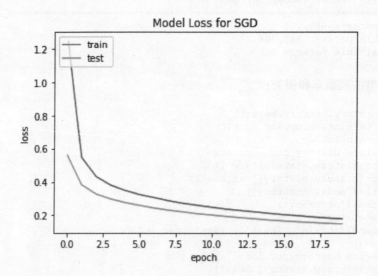

输出最终的准确率：

```
score = model.evaluate(x_test, y_test, verbose=0)
print('Test loss:', score[0])
print('Test accuracy:', score[1])
```

结果显示准确率达到 0.956，这已经是个很高的值，但以下小节中将会对其继续改进。

```
Test loss: 0.147237592921
Test accuracy: 0.956
```

3.7 Adam 优化算法

自适应矩估计（Adam） 计算每个参数的自适应学习率。与 AdaDelta 一样，Adam 存储过去平方梯度的衰减平均值和每个参数的动态变化。Adam 在实践中运作良好，是当今最常用的优化方法之一。

除了每一时刻平方梯度衰减的加权平均值（如 Adadelta 和 RMSprop）之外，Adam 还存储每一时刻梯度衰减指数 m_t 的加权平均值。使用以下公式计算 m_t 和 v_t：

$$m_t = \beta_1 m_{t-1} + (1-\beta_1) g_t$$
$$v_t = \beta_2 v_{t-1} + (1-\beta_2) g_t$$

m_t 和 v_t 分别是梯度中第一时刻（平均值）和第二时刻（未中心化方差）的估计值，在初始化时衰减率很小（即 β_1 和 β_2 接近 1），m_t 和 v_t 被初始化为零向量。

Adam 算法的设计者利用偏差校正第一时刻和第二时刻的估计值来抵消这些偏差，更新公式如下：

$$\hat{m} = \frac{m}{1-\beta_1^t}$$

$$\hat{v} = \frac{v}{1-\beta_2^t}$$

$$\Theta_{t+1} = \Theta_t - \frac{\eta}{\sqrt{\hat{v}_t + \epsilon}} \cdot \hat{m}_t$$

3.7.1 准备工作

在执行之前，需要在主代码段前添加前面的示例通用代码。

3.7.2 怎么做

使用适当网络拓扑创建一个序贯模型：

- 输入层：输入维度（*，784），输出维度（*，512）。
- 隐藏层：输入维度（*，512），输出维度（*，512）。

- 输出层：输入维度（*，512），输出维度（*，10）。

每层的激活函数如下所示：
- 第1层和第2层：relu 函数
- 第3层：softmax 函数

```
y_train = keras.utils.to_categorical(y_train, num_classes)
y_test = keras.utils.to_categorical(y_test, num_classes)

model = Sequential()
model.add(Dense(512, activation='relu', input_shape=(784,),))
model.add(Dropout(0.2))
model.add(Dense(512, activation='relu'))
model.add(Dropout(0.2))
model.add(Dense(num_classes, activation='softmax'))

model.summary()
adam = keras.optimizers.Adam(lr=0.001, beta_1=0.9, beta_2=0.999,
epsilon=None, decay=0.0, amsgrad=False)
model.compile(loss='categorical_crossentropy',
optimizer=adam,metrics=['accuracy'])


history = model.fit(x_train, y_train,
 batch_size=batch_size,
 epochs=epochs,
 verbose=1,
 validation_data=(x_test, y_test))
```

这里创建了一个具有两个隐藏层，丢弃率为 0.2 的网络。

上述代码的输出：

```
Layer (type) Output Shape Param #
=================================================================
dense_10 (Dense) (None, 512) 401920
_____
dropout_7 (Dropout) (None, 512) 0
_____
dense_11 (Dense) (None, 512) 262656
_____
dropout_8 (Dropout) (None, 512) 0
_____
dense_12 (Dense) (None, 10) 5130
=================================================================
Total params: 669,706
Trainable params: 669,706
Non-trainable params: 0
```

打印模型的准确率和损失：

```
print(history.history.keys())
import matplotlib.pyplot as plt
%matplotlib inline
# summarize history for accuracy
plt.plot(history.history['acc'])
plt.plot(history.history['val_acc'])
plt.title('model accuracy')
plt.ylabel('accuracy')
plt.xlabel('epoch')
plt.legend(['train', 'test'], loc='upper left')
plt.show()
# summarize history for loss
plt.plot(history.history['loss'])
plt.plot(history.history['val_loss'])
plt.title('model loss')
plt.ylabel('loss')
plt.xlabel('epoch')
plt.legend(['train', 'test'], loc='upper left')
plt.show()
```

测试数据和训练数据的模型准确率曲线如下图所示，均可收敛到 95%：

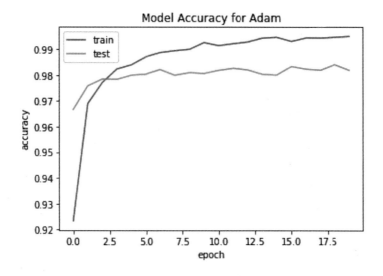

相应的模型损失曲线如下图所示：

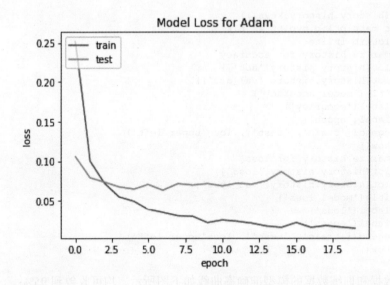

最终的准确率为：

```
score = model.evaluate(x_test, y_test, verbose=0)
print('Test loss:', score[0])
print('Test accuracy:', score[1])
```

结果显示准确率达到 0.982，远高于 SGD 的结果：

```
Test loss: 0.0721200712588
Test accuracy: 0.982
```

在第 3.8 节，我们还将介绍 AdaDelta 算法。

3.8 AdaDelta 优化算法

AdaDelta 解决了 AdaGrad 优化算法学习率下降的问题。AdaGrad 的学习率为 1 除以平方根的总和，每个阶段会添加一个平方根，使得分母不断增加。而 AdaDelta 不是对所有先前的平方根求和，而是使用允许总和减少的滑动窗口。

AdaDelta 是 AdaGrad 的改进，减缓了学习率的下降速率。AdaDelta 不是累积所有过去的平方梯度，而是将累积过去梯度的窗口限制为固定大小 w。

AdaDelta 不是低效地存储 w 大小的过去平方梯度，而是将梯度的总和递归地定义为所有过去的平方梯度的衰减平均值。时间步长 t 的运行平均值 $E[g^2]_t$ 仅依赖于先前的平均值和当前梯度（γ 作为系数，类似于动量项）：

$$E[g^2]_t = \gamma E[g^2]_{t-1} + (1-\gamma) g_t^2$$

其中 $E[g^2]_t$ 是时间 t 时的梯度的平方和，$\gamma E[g^2]_{t-1}$ 是时间 $t-1$ 的梯度平方和的 γ 倍，其中 γ 是 $E[g^2]_{t-1}$ 被添加到等式的系数。

假设 θ 有增量，则：

$$\Delta\theta_t = -\eta \cdot g_{t,i}$$
$$\theta_{t+1} = \theta_t + \Delta\theta_t$$

所以新的 $\Delta\theta$ 项为：

$$\Delta\theta_t = \frac{-\eta}{\sqrt{E[g^2]_t + \epsilon}}$$

3.8.1 准备工作

引入前面的示例通用代码指定的类、方法等。

3.8.2 怎么做

和前几节一样，利用适当的网络拓扑创建一个序贯模型。本节中使用的优化器是 Keras 中的 `AdaDelta` 实现：

```
keras.optimizers.Adadelta(lr=1.0, rho=0.95, epsilon=None, decay=0.0)
```

AdaDelta 优化器

Keras 文档中建议将此优化器的参数保留为默认值。

优化器的初始化参数：

- `lr`：大于等于 0 的浮点数，表示学习率。建议将其保留为默认值。
- `rho`：大于等于 0 的浮点数。
- `epsilon`：大于等于 0 的浮点数，表示模糊因子。如果未指定（`None`），则默认为 `K.epsilon()`。
- `decay`：大于等于 0 的浮点数，表示每次数据更新时，衰减的学习率。

```
model = Sequential()
model.add(Dense(512, activation='relu', input_shape=(784,)))
model.add(Dropout(0.2))
model.add(Dense(512, activation='relu'))
model.add(Dropout(0.2))
model.add(Dense(num_classes, activation='softmax'))
model.summary()
ada_delta = keras.optimizers.Adadelta(lr=1.0, rho=0.95,
 epsilon=None, decay=0.0)
model.compile(loss='categorical_crossentropy',
 optimizer=ada_delta,
 metrics=['accuracy'])

history = model.fit(x_train, y_train,
 batch_size=batch_size,
 epochs=epochs,
 verbose=1,
 validation_data=(x_test, y_test))
```

这里创建了一个具有两个隐藏层，丢弃率为 0.2 的网络。

该模型使用的是 AdaDelta 优化器。

以下是上述代码的输出：

```
Layer (type) Output Shape Param #
=================================================================
dense_1 (Dense) (None, 512) 401920
_____
dropout_1 (Dropout) (None, 512) 0
_____
dense_2 (Dense) (None, 512) 262656
_____
dropout_2 (Dropout) (None, 512) 0
_____
dense_3 (Dense) (None, 10) 5130
=================================================================
Total params: 669,706
Trainable params: 669,706
Non-trainable params: 0
```

绘制 AdaDelta 的模型准确率曲线：

```
import matplotlib.pyplot as plt
%matplotlib inline
# summarize history for accuracy
plt.plot(history.history['acc'])
plt.plot(history.history['val_acc'])
plt.title('Model Accuracy for RMSProp')
plt.ylabel('accuracy')
plt.xlabel('epoch')
plt.legend(['train', 'test'], loc='upper left')
```

```
plt.show()
# summarize history for loss
plt.plot(history.history['loss'])
plt.plot(history.history['val_loss'])
plt.title('Model Loss for RMSProp')
plt.ylabel('loss')
plt.xlabel('epoch')
plt.legend(['train', 'test'], loc='upper left')
plt.show()
```

基于 AdaDelta 优化器的训练集准确率和测试集准确率：

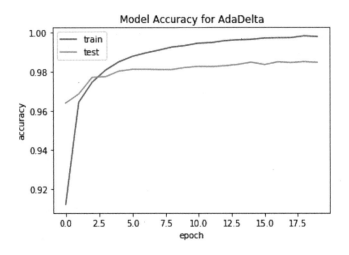

同样，AdaDelta 的模型损失曲线如下所示：

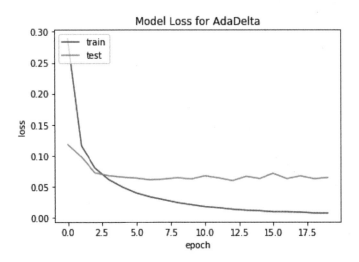

AdaDelta 的最终测试损失和测试准确率计算如下：

```
score = model.evaluate(x_test, y_test, verbose=0)
print('Test loss:', score[0])
print('Test accuracy:', score[1])
```

上述程序的输出显示如下：

```
Test loss: 0.0644025499775
Test accuracy: 0.9846
```

使用 AdaDelta 获得的准确率高于 SGD、Adam 的结果，大约为 0.9846。

3.9 使用 RMSProp 进行优化

本节将介绍使用 RMSProp 进行优化的相关示例代码。

RMSProp 是由 Geoff Hinton 提出的（未发表的）自适应学习方法。RMSProp 和 AdaDelta 是在同一时期独立开发的，其目的都是为了解决 AdaGrad 中学习率急剧下降的问题。

RMSProp 中的第一个更新向量与之前介绍的 AdaDelta 相同：

$$E\left[g^2\right]_t = 0.9 E\left[g^2\right]_{t-1} + 0.1 g_t^2$$

$$\Theta_{t+1} = \Theta_t - \frac{\eta}{\sqrt{E\left[g^2\right]_t + \epsilon}} \cdot g_t$$

RMSProp 算法将学习率除以平方梯度的指数衰减平均值。建议将 γ 设定为 0.9，学习率 η 设置为 0.001。

3.9.1 准备工作

引入前面的示例通用代码指定的类、方法等。

3.9.2 怎么做

创建序贯模型：

```
from keras.optimizers import RMSprop
model = Sequential()
```

```
model.add(Dense(512, activation='relu', input_shape=(784,)))
model.add(Dropout(0.2))
model.add(Dense(512, activation='relu'))
model.add(Dropout(0.2))
model.add(Dense(num_classes, activation='softmax'))
model.summary()
model.compile(loss='categorical_crossentropy',
 optimizer=RMSprop(),
 metrics=['accuracy'])
history = model.fit(x_train, y_train,
 batch_size=batch_size,
 epochs=epochs,
 verbose=1,
 validation_data=(x_test, y_test))
```

这里创建了一个具有两个隐藏层、丢弃率为 0.2 的网络。

使用的优化器为 RMSProp。

以下是上述代码的输出：

```
Layer (type) Output Shape Param #
=================================================================
dense_1 (Dense) (None, 512) 401920
_____
dropout_1 (Dropout) (None, 512) 0
_____
dense_2 (Dense) (None, 512) 262656
_____
dropout_2 (Dropout) (None, 512) 0
_____
dense_3 (Dense) (None, 10) 5130
=================================================================
Total params: 669,706
Trainable params: 669,706
Non-trainable params: 0
```

绘制 RMSProp 的模型准确率曲线：

```
import matplotlib.pyplot as plt
%matplotlib inline
# summarize history for accuracy
plt.plot(history.history['acc'])
plt.plot(history.history['val_acc'])
plt.title('Model Accuracy for RMSProp')
plt.ylabel('accuracy')
plt.xlabel('epoch')
plt.legend(['train', 'test'], loc='upper left')
plt.show()
# summarize history for loss
plt.plot(history.history['loss'])
plt.plot(history.history['val_loss'])
```

```
plt.title('Model Loss for RMSProp')
plt.ylabel('loss')
plt.xlabel('epoch')
plt.legend(['train', 'test'], loc='upper left')
plt.show()
```

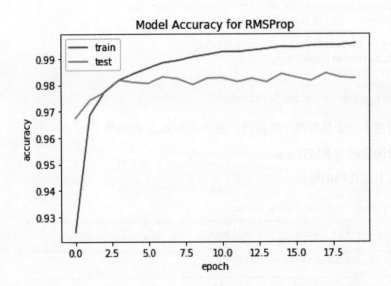

同样，模型损失曲线如下图所示：

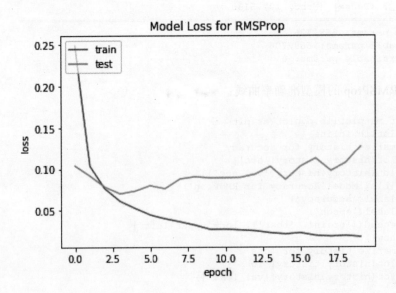

RMSProp 的最终测试损失和测试准确率计算如下：

```
score = model.evaluate(x_test, y_test, verbose=0)
print('Test loss:', score[0])
print('Test accuracy:', score[1])
```

输出如下：

```
Test loss: 0.126795965524
Test accuracy: 0.9824
```

使用 RMSProp 获得的准确率约为 0.95，高于普通 SGD，但低于 Adam 和 AdaDelta 的结果。

CHAPTER 4

第 4 章

使用不同的 Keras 层实现分类

本章包括以下内容：
- 乳腺癌分类
- 垃圾信息检测分类

4.1 引言

分类是一个经典问题，其目的是将语料库或图像自动分为一个或多个指定的类，本章将介绍如何利用 Keras 库来训练分类器。分类是机器学习过程中的常见任务，例如问答、支持请求、业务问题、癌症分类等。本章介绍了如何利用开源神经网络库解决分类问题。深度学习具有更灵活的模型，需要较少的专业领域知识，以及更容易进行持续学习。所以与传统算法相比，深度学习在分类方面更具优势。

4.2 乳腺癌分类

我们的分类问题的目的是预测癌症是处于良性期还是恶性期。通过开发基于神经网络的算法来准确预测乳腺癌是良性还是恶性（准确率约为 94%），就像教一台机器去预测乳腺癌一样。我们将使用 Keras API 来构建，这看似复杂的模型。以下内容提供了数据集描述：

> **威斯康星乳腺癌（诊断）数据集：**
> 特征从乳房肿块的**细针抽吸活检（FNA）**数字化图像中计算得到，这些值代表

出现在图像中的细胞核的特征。部分图像可以在 http://www.cs.wisc.edu/~street/images/ 找到。

分类分布：357 良性，212 恶性。

该数据库也可通过 UW CS FTP 服务器获得：ftp ftp.cs.wisc.edu cd math-prog/cpo-dataset/machine- learn/WDBC/

怎么做

本节将开发一个分类模型流水线以分类癌症类型。模型流水线使用基于 Keras 函数 API 的 Adam 模型，也会调用各种数据操作库。我们会使用 Keras，具体地说，我们将使用 `sklearn`、`numpy` 和 `pandas` 来创建模型。

首先导入以下库：`pandas` 是 Python 中的数据分析库；`sklearn` 是一个以数据挖掘和分析而闻名的机器学习库；Keras 是一个神经网络 API，将帮助我们创建实际的神经网络。

```
import pandas as pd
from sklearn.preprocessing import LabelEncoder, StandardScaler
from sklearn.model_selection import train_test_split
from sklearn.model_selection import GridSearchCV
from keras.wrappers.scikit_learn import KerasClassifier
from keras.models import Sequential, load_model
from keras.layers import Dense
from sklearn.metrics import confusion_matrix
```

数据处理

在神经网络进行训练和验证的过程中，输入数据的正确性对最终结果有着巨大的影响。为了保证结果的一致性和正确性，需要确保数据具有合适的比例和格式，而且必须包含有意义的特征。

执行以下步骤进行数据预处理：

1. 使用 `pandas` 加载数据集
2. 将数据集拆分为输入和输出变量以进行机器学习
3. 对输入变量进行预处理变换
4. 将数据汇总以显示其变化

使用 pandas 库来加载数据并查看数据集的形状：

```
dataset = pd.read_csv('/deeplearning/google/kaggle/breast-cancer/data.csv')

# get dataset details
print(dataset.head(5))
print(dataset.columns.values)
print(dataset.info())
print(dataset.describe())
```

输出为如下代码所示：

```
   id diagnosis    ...    fractal_dimension_worst  Unnamed: 32
0     842302    M  ...                    0.11890          NaN
1     842517    M  ...                    0.08902          NaN
2   84300903    M  ...                    0.08758          NaN
3   84348301    M  ...                    0.17300          NaN
4   84358402    M  ...                    0.07678          NaN
Data columns (total 33 columns):
id                         569 non-null int64
diagnosis                  569 non-null object
radius_mean                569 non-null float64
texture_mean               569 non-null float64
perimeter_mean             569 non-null float64
area_mean                  569 non-null float64
smoothness_mean            569 non-null float64
compactness_mean           569 non-null float64
concavity_mean             569 non-null float64
concave points_mean        569 non-null float64
symmetry_mean              569 non-null float64
fractal_dimension_mean     569 non-null float64
radius_se                  569 non-null float64
texture_se                 569 non-null float64
perimeter_se               569 non-null float64
area_se                    569 non-null float64
smoothness_se              569 non-null float64
compactness_se             569 non-null float64
concavity_se               569 non-null float64
concave points_se          569 non-null float64
symmetry_se                569 non-null float64
fractal_dimension_se       569 non-null float64
radius_worst               569 non-null float64
texture_worst              569 non-null float64
perimeter_worst            569 non-null float64
area_worst                 569 non-null float64
smoothness_worst           569 non-null float64
compactness_worst          569 non-null float64
concavity_worst            569 non-null float64
concave points_worst       569 non-null float64
symmetry_worst             569 non-null float64
fractal_dimension_worst    569 non-null float64
```

```
Unnamed: 32              0 non-null float64
dtypes: float64(31), int64(1), object(1)
memory usage: 146.8+ KB
None
        id          ...   Unnamed: 32
count   5.690000e+02 ...           0.0
mean    3.037183e+07 ...           NaN
std     1.250206e+08 ...           NaN
min     8.670000e+03 ...           NaN
25%     8.692180e+05 ...           NaN
50%     9.060240e+05 ...           NaN
75%     8.813129e+06 ...           NaN
max     9.113205e+08 ...           NaN
```

输出不仅显示了数据集的起始值、特征值、缺失值,还包括了数字特征的分布。

为了使我们的模型能够预测肿瘤是恶性还是良性,将已分类的肿瘤数据集作为输入。数据集共有 33 列,排除掉未命名且全为空值的最后一列,有意义的有 32 列。其中,`diagnosis` 是标签或类别,而 `id` 对分类不造成什么影响,所以这两列也不包括在训练集中。因此,保留下来类型为 `float 64` 且不包含缺失值的 30 列特征。

拆分特征和标签:

```
# data cleansing
X = dataset.iloc[:, 2:32]
print(X.info())
print(type(X))
y = dataset.iloc[:, 1]
print(y)
```

输出如下:

```
Data columns (total 30 columns):
radius_mean               569 non-null float64
texture_mean              569 non-null float64
...
concave points_worst      569 non-null float64
symmetry_worst            569 non-null float64
fractal_dimension_worst   569 non-null float64
dtypes: float64(30)
memory usage: 133.4 KB
None
<class 'pandas.core.frame.DataFrame'>
0    M
1    M
2    M
3    M
4    M
```

`diagnosis` 被标记为 M 或 B 以代表恶性或良性肿瘤，继续使用 `LabelEncoder()` 将它们编码为 0 和 1：

```
'''encode the labels to 0, 1 respectively'''
print(y[100:110])
encoder = LabelEncoder()
y = encoder.fit_transform(y)
print([y[100:110]])
```

输出如下：

```
Name: diagnosis, dtype: object
[array([1, 0, 0, 0, 0, 1, 0, 0, 1, 0])]
```

现在，将数据分成训练集和验证集。80% 的数据作为训练集用于拟合模型；20% 的数据作为验证集，用于对训练集拟合的模型进行无偏评估，并调优超参数。

```
# lets split dataset now
XTrain, XTest, yTrain, yTest = train_test_split(X, y, test_size=0.2, random_state=0)
```

现在进行特征缩放，避免因为有的特征过大，导致模型无法将它们用作主要预测因子：

```
# feature scaling
scalar = StandardScaler()
XTrain = scalar.fit_transform(XTrain)
XTest = scalar.transform(XTest)
```

建模

创建一个模型并添加层。我们可以对模型单元数进行初始化修改，如果并不确定要初始化的数量，那么只需对除了最后一个层（最后一个层使用特征的数量与输出节点的数量求平均，本例中为 15）以外的所有层单元进行初始化处理。首先我们必须为第一个层提供一个输入维度。激活方式中 `relu` 是指整流线性单元激活，`sigmoid` 指的是 `sigmoid` 激活函数。在 `sigmoid` 激活函数的帮助下，我们可以得到分类的概率，这在某些情况下有益于进一步的研究。

对于每个模型，都有一个默认的超参数。首先需要找到能使模型给出更精确预测的超参数。这里我们需要花一些时间来调优 `batch_size`、`epochs` 和 `optimizer`。

调优是机器学习流水线中显示结果之前的最后一步，也被称为**超参数优化**，其中算法的参数被看作超参数,而系数被看作参数。优化意在提高问题的搜索能力。网格搜索是一种参数调优的方法,它将为网格中指定的各种算法参数的构建并评估模型。

1. **GridSearchCV** 可以帮助估算器对指定参数值进行搜索:

```
# choosing hyper parameters
def classifier(optimizer):
    model = Sequential()
    model.add(Dense(units=16, kernel_initializer='uniform', activation='relu', input_dim=30))
    model.add(Dense(units=8, kernel_initializer='uniform', activation='relu'))
    model.add(Dense(units=6, kernel_initializer='uniform', activation='relu'))
    model.add(Dense(units=1, kernel_initializer='uniform', activation='sigmoid'))
    model.compile(optimizer=optimizer, loss='binary_crossentropy', metrics=['accuracy'])
    return model


model = KerasClassifier(build_fn=classifier)
params = {'batch_size': [1, 5], 'epochs': [100, 120], 'optimizer': ['adam', 'rmsprop']}
gridSearch = GridSearchCV(estimator=model, param_grid=params, scoring='accuracy', cv=10)
gridSearch = gridSearch.fit(XTrain, yTrain)
score = gridSearch.best_score_
bestParams = gridSearch.best_params_
print(score)
print(bestParams)
```

输出如下:

```
1/455 [..............................] - ETA: 3:36 - loss: 0.6932 - acc: 0.0000e+00
44/455 [=>............................] - ETA: 4s - loss: 0.6928 - acc: 0.5227
88/455 [====>.........................] - ETA: 2s - loss: 0.6912 - acc: 0.6136
130/455 [=======>......................] - ETA: 1s - loss: 0.6849 - acc: 0.7154
166/455 [=========>....................] - ETA: 1s - loss: 0.6723 - acc: 0.7530
200/455 [=============>................] - ETA: 0s - loss: 0.6613 - acc: 0.7850
```

best_parameters: {'batch_size': 1, 'epochs': 100, 'optimizer': 'rmsprop'}
best_accuracy: 0.998021978022

2. 用找到的参数构建神经网络：

```
# modeling
model = Sequential()
model.add(Dense(units=16, kernel_initializer='uniform',
 activation='relu', input_dim=30))
model.add(Dense(units=8, kernel_initializer='uniform',
 activation='relu'))
model.add(Dense(units=6, kernel_initializer='uniform',
 activation='relu'))
model.add(Dense(units=1, kernel_initializer='uniform',
 activation='sigmoid'))
```

3. 使用 adam 优化器编译分类器并使用 binary_crossentropy 作为损失函数，因为这是一个二元分类，即只有两个类：M 或 B。

```
model.compile(optimizer='adam', loss='binary_crossentropy',
 metrics=['accuracy'])
```

4. 拟合数据，以批次大小为 1、迭代次数为 120 进行训练，并保存模型以供以后的分类任务使用：

```
model.fit(XTrain, yTrain, batch_size=1, epochs=120)
model.save('./cancer_model.h5')
```

5. 用该模型对数据集进行分类和测试，计算训练准确度和混淆矩阵：

```
yPred = model.predict(XTest)
yPred = [1 if y > 0.5 else 0 for y in yPred]
matrix = confusion_matrix(yTest, yPred)
print(matrix)
accuracy = (matrix[0][0] + matrix[1][1]) / (matrix[0][0] +
matrix[0][1] + matrix[1][0] + matrix[1][1])
print("Accuracy: " + str(accuracy * 100) + "%")
```

输出显示如下：

```
[[64  3]
 [ 3 44]]
Accuracy: 94.736842105263155%
```

完整代码

以下是乳腺癌分类的完整代码：

```
import pandas as pd
from sklearn.preprocessing import LabelEncoder, StandardScaler
```

```python
from sklearn.model_selection import train_test_split
from sklearn.model_selection import GridSearchCV
from keras.wrappers.scikit_learn import KerasClassifier
from keras.models import Sequential, load_model
from keras.layers import Dense
from sklearn.metrics import confusion_matrix

dataset = pd.read_csv('/deeplearning/google/kaggle/breast-cancer/data.csv')

# get dataset details
print(dataset.head(5))
print(dataset.columns.values)
print(dataset.info())
print(dataset.describe())

# data cleansing
X = dataset.iloc[:, 2:32]
print(X.info())
print(type(X))
y = dataset.iloc[:, 1]
print(y)

'''encode the labels to 0, 1 respectively'''
print(y[100:110])
encoder = LabelEncoder()
y = encoder.fit_transform(y)
print([y[100:110]])

# lets split dataset now
XTrain, XTest, yTrain, yTest = train_test_split(X, y, test_size=0.2,
random_state=0)

# feature scaling
scalar = StandardScaler()
XTrain = scalar.fit_transform(XTrain)
XTest = scalar.transform(XTest)

# choosing hyper parameters
'''
def classifier(optimizer):
    model = Sequential()
    model.add(Dense(units=16, kernel_initializer='uniform',
activation='relu', input_dim=30))
    model.add(Dense(units=8, kernel_initializer='uniform',
activation='relu'))
    model.add(Dense(units=6, kernel_initializer='uniform',
activation='relu'))
    model.add(Dense(units=1, kernel_initializer='uniform',
activation='sigmoid'))
    model.compile(optimizer=optimizer, loss='binary_crossentropy',
metrics=['accuracy'])
    return model
```

```
model = KerasClassifier(build_fn=classifier)
params = {'batch_size': [1, 5], 'epochs': [100, 120], 'optimizer': ['adam',
'rmsprop']}
gridSearch = GridSearchCV(estimator=model, param_grid=params,
scoring='accuracy', cv=10)
gridSearch = gridSearch.fit(XTrain, yTrain)
score = gridSearch.best_score_
bestParams = gridSearch.best_params_
print(score)
print(bestParams)
'''

# modeling
model = Sequential()
model.add(Dense(units=16, kernel_initializer='uniform', activation='relu',
input_dim=30))
model.add(Dense(units=8, kernel_initializer='uniform', activation='relu'))
model.add(Dense(units=6, kernel_initializer='uniform', activation='relu'))
model.add(Dense(units=1, kernel_initializer='uniform',
activation='sigmoid'))
model.compile(optimizer='adam', loss='binary_crossentropy',
metrics=['accuracy'])
model.fit(XTrain, yTrain, batch_size=1, epochs=120)
model.save('/Users/manpreet.singh/git/deeplearning/google/kaggle/breast-
cancer/cancer_model.h5')
yPred = model.predict(XTest)
yPred = [1 if y > 0.5 else 0 for y in yPred]
matrix = confusion_matrix(yTest, yPred)
print(matrix)
accuracy = (matrix[0][0] + matrix[1][1]) / (matrix[0][0] + matrix[0][1] +
matrix[1][0] + matrix[1][1])
print("Accuracy: " + str(accuracy * 100) + "%")
```

4.3 垃圾信息检测分类

垃圾信息检测是一种常见的分类问题。以下小节中提到的原始文本和原始文档的语料库中，包括标记为垃圾信息和非垃圾信息的文档。数据源是 SMS Spam Collection v.1，它是针对手机垃圾信息研究而收集的一组公共 SMS 标记信息。

数据集可以从 http://www.dt.fee.unicamp.br/~tiago/smsspamcollection/ 下载。下表列出了数据集的不同文件格式，以及每个类中的样本数和样本总数：

应用	文件格式	垃圾信息	正常信息	总数	链接
通用	纯文本	747	4,827	5,574	http://www.dt.fee.unicamp.br/~tiago/ smsspamcollection/smsspamcollection.zip
Weka	ARFF	747	4,827	5,574	http://www.dt.fee.unicamp.br/~tiago/ smsspamcollection/smsSpamCollection.arff

怎么做

本小节中,我们开发一个分类模型流水线,将信息类型分为正常信息和垃圾信息,模型流水线使用 Keras 函数 API 编写的 RMSProp 模型。

导入以下库以供使用。`pandas` 是 Python 中的数据分析库,`numpy` 是一个数值计算库,`keras` 是一个可以帮助创建实际神经网络的神经网络 API:

```
from keras.layers import SimpleRNN, Embedding, Dense, LSTM
from keras.models import Sequential
from keras.preprocessing.text import Tokenizer
from keras.preprocessing.sequence import pad_sequences
import numpy as np
from sklearn.metrics import confusion_matrix
import matplotlib.pyplot as plt
import pandas as pd
```

数据处理

我们将使用 `pandas` 库来加载数据并查看数据集的形状。

执行以下步骤以进行数据预处理:

1. 使用 `pandas` 加载数据集

2. 将标签转换为 `[0,1]`

3. 将数据集拆分为输入和输出变量以进行机器学习

4. 令牌化数据

5. 输出数据信息以发现对数据的改动情况:

```
# get dataset
data = pd.read_csv('./data.csv')
texts = []
classes = []
for i, label in enumerate(data['Class']):
    texts.append(data['Text'][i])
    if label == 'ham':
        classes.append(0)
    else:
        classes.append(1)

texts = np.asarray(texts)
classes = np.asarray(classes)

print("number of texts :", len(texts))
print("number of labels: ", len(classes))
```

输出如下：

```
number of texts: 5572
number of labels: 5572
```

6.定义分类器所用的特征和文档长度的最大值：

```
# number of words used as features
maxFeatures = 10000
# max document length
maxLen = 500
```

7.将数据分成训练集和验证集。80%的数据作为训练集用于拟合模型；20%的数据作为验证集，用于对训练集拟合的模型进行无偏评估，并调优超参数。

```
# we will use 80% of data as training and 20% as validation data
trainingData = int(len(texts) * .8)
validationData = int(len(texts) - trainingData)
```

8.词被视为**令牌**，将文本分解为令牌的方法被称为**令牌化**。Keras库提供了`Tokenizer()`类，用于为神经网络准备文本文档。创建`tokenizer`并在原始文本文档或整数编码文本文档进行拟合：

```
# tokenizer
tokenizer = Tokenizer()
tokenizer.fit_on_texts(texts)
sequences = tokenizer.texts_to_sequences(texts)
word_index = tokenizer.word_index
print("Found {0} unique words: ".format(len(word_index)))
data = pad_sequences(sequences, maxlen=maxLen)
print("data shape: ", data.shape)
```

输出显示在以下代码中：

```
Found 9006 unique words:
data shape: (5572, 500)
```

9.最后，我们将数据集混合并创建用于建模的训练集和测试集，如下所示：

```
# shuffle data
indices = np.arange(data.shape[0])
np.random.shuffle(indices)
data = data[indices]
labels = classes[indices]

X_train = data[:trainingData]
y_train = labels[:trainingData]
```

```
X_test = data[trainingData:]
y_test = labels[trainingData:]
```

建模

现在使用 Keras 库创建一个序贯模型,该模型内部为一系列层。首先,我们创建一个新的序贯模型并对其添加层来构建网络拓扑。定义模型后,利用 TensorFlow 作为后端进行编译。此后端选择基于以最佳方式在给定硬件上运行,对网络进行训练并进行预测。

在网络建模的过程中加入嵌入层,如以下代码片段所示。嵌入层的最大特征数为 32。此时构建出的模型是一个二元分类器。最后,可以对分类模型进行拟合、评估。

我们需要指定损失函数来估计权重值,用优化器对不同权重进行搜索,并收集我们希望在训练期间获得的可选指标。代码如下:

```
# modeling
model = Sequential()
model.add(Embedding(maxFeatures, 32))
model.add(LSTM(32))
model.add(Dense(1, activation='sigmoid'))
model.compile(optimizer='rmsprop', loss='binary_crossentropy',
metrics=['acc'])
rnn = model.fit(X_train, y_train, epochs=10, batch_size=60,
validation_split=0.2)
```

输出如下:

```
Epoch 1/10
60/3565 [..........................] - ETA: 2:00 - loss: 0.6927 - acc:
0.5833
 120/3565 [>.........................] - ETA: 1:25 - loss: 0.6864 -
acc: 0.7083
 180/3565 [>.........................] - ETA: 1:13 - loss: 0.6812 -
acc: 0.7556
...
Epoch 10/10
3000/3565 [========================>.....] - ETA: 9s - loss: 0.0175 - acc:
0.9933
 3540/3565 [=============================>.] - ETA: 0s - loss: 0.0155 - acc:
0.9944
 3565/3565 [==============================] - 61s 17ms/step - loss: 0.0154
- acc: 0.9944 - val_loss: 0.0463 - val_acc: 0.9865
```

最后,在测试数据上对模型进行性能评估:

```
# predictions
pred = model.predict_classes(X_test)
```

```
acc = model.evaluate(X_test, y_test)
proba_rnn = model.predict_proba(X_test)
print("Test loss is {0:.2f} accuracy is {1:.2f}
".format(acc[0],acc[1]))
print(confusion_matrix(pred, y_test))
```

输出如下：

```
Test loss is 0.07 accuracy is 0.98
[[956 15]
 [  5 139]]
```

训练集和验证集的模型损失如下图所示：

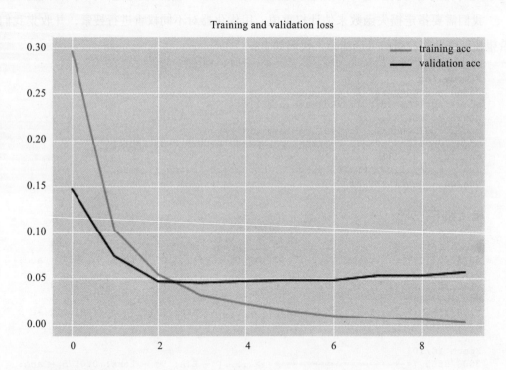

完整代码

以下是垃圾信息检测分类的完整代码：

```
from keras.layers import Embedding, Dense, LSTM
from keras.models import Sequential
from keras.preprocessing.text import Tokenizer
from keras.preprocessing.sequence import pad_sequences
import numpy as np
from sklearn.metrics import confusion_matrix
```

```python
import pandas as pd

# get dataset
data = pd.read_csv('/spam-detection/spam_dataset.csv')
texts = []
classes = []
for i, label in enumerate(data['Class']):
    texts.append(data['Text'][i])
    if label == 'ham':
        classes.append(0)
    else:
        classes.append(1)

texts = np.asarray(texts)
classes = np.asarray(classes)

print("number of texts :", len(texts))
print("number of labels: ", len(classes))

# number of words used as features
maxFeatures = 10000
# max document length
maxLen = 500

# we will use 80% data set for training and 20% data set for validation
trainingData = int(len(texts) * .8)
validationData = int(len(texts) - trainingData)

# tokenizer
tokenizer = Tokenizer()
tokenizer.fit_on_texts(texts)
sequences = tokenizer.texts_to_sequences(texts)
word_index = tokenizer.word_index
print("Found {0} unique words: ".format(len(word_index)))
data = pad_sequences(sequences, maxlen=maxLen)
print("data shape: ", data.shape)

indices = np.arange(data.shape[0])
np.random.shuffle(indices)
data = data[indices]
labels = classes[indices]

X_train = data[:trainingData]
y_train = labels[:trainingData]
X_test = data[trainingData:]
y_test = labels[trainingData:]

# modeling
model = Sequential()
model.add(Embedding(maxFeatures, 32))
model.add(LSTM(32))
model.add(Dense(1, activation='sigmoid'))
model.compile(optimizer='rmsprop', loss='binary_crossentropy',
```

```
metrics=['acc'])
rnn = model.fit(X_train, y_train, epochs=10, batch_size=60,
validation_split=0.2)

# predictions
pred = model.predict_classes(X_test)
acc = model.evaluate(X_test, y_test)
proba_rnn = model.predict_proba(X_test)
print("Test loss is {0:.2f} accuracy is {1:.2f}  ".format(acc[0],acc[1]))
print(confusion_matrix(pred, y_test))
```

CHAPTER 5

第 5 章

卷积神经网络的实现

本章将介绍以下内容：
- 宫颈癌分类
- 数字识别

5.1 引言

卷积神经网络（CNN） 是具有可学习权重和偏差的神经元的网络。每个神经元接受输入，计算点积，然后进行非线性运算。CNN 包含几个卷积层，然后是一个或多个全连接的层，如在标准多层神经网络中，从接收原始图像像素开始到最后的类别分数。CNN 通过学习特征表示来保持像素之间的空间关系。该特征被学习并应用于整个图像，允许图像中的对象在场景中旋转或平移时仍然能被网络检测到。

简而言之，CNN 基本上是由几个卷积层以及非线性激活函数如 ReLU 或 tanh 构成，最后生成结果。

CNN 的应用包括关系提取和关系分类任务、图像处理和自然语言处理，可能使用 CNN 的情景更多。本章的内容旨在向你介绍应用于深度学习模型的 CNN，以便你可以轻松地将它应用于你的数据集并开发出有用的程序。

5.2 宫颈癌分类

宫颈癌是发生在子宫颈的癌症。如果宫颈癌在早期发现很容易治愈。然而，由于该领域缺乏专业知识,宫颈癌筛查和治疗计划面临的最大挑战之一是确定合适的治疗方法,而如果能够实时确定宫颈癌的类型,治疗工作将得到极大提高。

5.2.1 准备工作

本节我们将开发一个模型流水线,基于图像以较高的准确率识别女性的宫颈类型,模型流水线使用以用于图像分类的 Keras 函数 API 编写的 CNN 模型,还会使用各种图像操作库。

> 该案例的数据可在 https://www.kaggle.com/c/intel-mobileodt-cervical-cancer-screening 中找到。该数据集是开发基于图像准确识别女性子宫颈类型的算法挑战的一部分。这样可以防止无效治疗,并允许医者为需要更高级治疗的病例提供适当的转诊。

首先,在 https://github.com/ml-resources/deeplearning-keras/tree/ed1/ch05 中下载 GitHub 库。将训练集和测试集下载并保存到 `data` 文件夹。

- `train.7z`:训练集。图像按其标记的类别进行组织,分为 `Type_1`、`Type_2` 和 `Type_3`。
- `test.7z`:测试集。

> 要了解有关如何定义这些子宫颈类型的更多背景信息,请参阅此文档:https://kaggle2.blob.core.windows.net/competitions/kaggle/6243/media/Cervix%20types%20clasification.pdf。

5.2.2 怎么做

将训练图像存储在 `train` 文件夹下的子文件夹 `Type_1`、`Type_2` 和 `Type_3` 中。存储格式为 `.jpg` 文件,并根据不同类别存储在对应的文件夹中。通过字符串读取图像路径,函数用图像的文件夹名称来获取相应标签,并将图像路径和标签作为两个并行的数组返回。接下来介绍数据预处理步骤。

数据预处理

宫颈图像具有不同的大小,且具有高分辨率。对于 CNN,输入数据不仅需要统一的大小，并且还需要足够的分辨率以便区分分类中的主要特征，而较低的分辨率可以避免计算能力上的限制。

```python
# process cervical dataset
def processCervicalData():
    # image resizing
    imgPaths = []
    labels = []
    trainingDirs = ['/deeplearning-keras/ch05/data/train']
    for dir in trainingDirs:
        newFilePaths, newLabels, numLabels = readFilePaths(dir)
        if len(newFilePaths) > 0:
            imgPaths += newFilePaths
            labels += newLabels

    imgPaths, labels = shuffle(imgPaths, labels)
    labelCount = labelsCount(labels)

    type1Count = labelCount[0]
    type2Count = labelCount[1]
    type3Count = labelCount[2]

    print("Count of type1 : ", type1Count)
    print("Count of type2 : ", type2Count)
    print("Count of type3 : ", type3Count)
    print("Total Number of data samples: " + str(len(imgPaths)))
    print("Number of Classes: " + str(numLabels))

    newShape = [(256,256,3)]
    destDir = ['/deeplearning-keras/ch05/data/resized_imgs']

    for newImgShape, destFolder in zip(newShape,destDir):
        for i, path,label in zip(count(),imgPaths,labels):
            split_path = path.split('/')
            newPath = 'size'+str(newImgShape[0])+'_'+split_path[-1]
            newPath = '/'.join([destFolder]+split_path[8:-1]+[newPath])
            add_flip = True
            if label == 1:
                add_flip = False

            # Used to exclude corrupt data
            try:
                resizeImage(path, maxSize=newImgShape, savePath=newPath, addFlip=add_flip)
            except OSError:
                print("Error at path " + path)
```

输出如下所示：

```
Using TensorFlow backend.
('Count of type1 : ', 250)
```

```
('Count of type2 : ', 781)
('Count of type3 : ', 450)
Total Number of data samples: 1481
Number of Classes: 3
```

1. 神经网络一方面需要图像的大小恒定,另一方面需要图像尽量小以留出足够的内存空间进行模型训练,但图像需要足够的像素以根据重要特征进行分类。我们半自动地选择了 $256×256$ 像素的图像大小。

2. 利用以下代码对图像大小进行等比例调整:

```
# Image resizing is important considering memory footprint, but its 
important to maintain key #characteristics that will preserve the 
key features.
def resizeImage(imgPath, maxSize=(256,256,3), savePath=None, 
addFlip=False):
    ImageFile.LOAD_TRUNCATED_IMAGES = True
    img = Image.open(imgPath)

    # set aspect ratio
    if type(img) == type(np.array([])):
        img = Image.fromarray(img)
    img.thumbnail(maxSize, Image.ANTIALIAS)
    tmpImage = (np.random.random(maxSize)*255).astype(np.uint8)
    resizedImg = Image.fromarray(tmpImage)
    resizedImg.paste(img,((maxSize[0]-img.size[0])//2, (maxSize[1]-
img.size[1])//2))
    if savePath:
        resizedImg.save(savePath)

    if addFlip:
        flip = resizedImg.transpose(Image.FLIP_LEFT_RIGHT)
        if savePath:
            splitPath = savePath.split('/')
            flip_path = '/'.join(splitPath[:-1] + 
['flipped_'+splitPath[-1]])
            flip.save(flip_path)
        return np.array(resizedImg, dtype=np.float32), 
np.array(flip,dtype=np.float32)
    return np.array(resizedImg, dtype=np.float32)
```

已调整的图像存储在 `resized_imgs` 文件夹中,具体如屏幕截图所示:

```
▼ 📁 resized_imgs
  ▼ 📁 train
    ▶ 📁 Type_1
    ▶ 📁 Type_2
    ▼ 📁 Type_3
        📄 flipped_size256_3.jpg
        📄 flipped_size256_5.jpg
        📄 flipped_size256_11.jpg
        📄 flipped_size256_16.jpg
```

3. 利用训练集对 CNN 进行训练。作为宫颈癌模型训练的一部分，读入调整后的图像路径，用于该项目的其余部分：

```
resizedImageDir = ['/deeplearning-
keras/ch05/data/resized_imgs/train']

imagePaths = []
labels = []
for i, resizedPath in enumerate(resizedImageDir):
    new_paths, new_labels, n_classes = readFilePaths(resizedPath)
    if len(new_paths) > 0:
        imagePaths += new_paths
        labels += new_labels

imagePaths, labels = shuffle(imagePaths, labels)
```

4. 将数据分成训练集和验证集。80% 的数据作为训练集用于拟合模型；20% 的数据组成验证集，用于模型超参数调整的同时提供训练集拟合模型的无偏评估。拆分数据后，将路径和标签保存为 CSV 文件，并注明它们所属的数据集，以确保两个数据集中的数据不会混淆：

```
trainCSV = '/deeplearning-keras/ch05/csvs/train_set.csv'
validCSV = '/deeplearning-keras/ch05/csvs/valid_set.csv'

training_portion = .8
split_index = int(training_portion * len(imagePaths))
X_train_paths, y_train = imagePaths[:split_index],
labels[:split_index]
X_valid_paths, y_valid = imagePaths[split_index:],
labels[split_index:]

print("Train size: ")
print(len(X_train_paths))
print("Valid size: ")
print(len(X_valid_paths))

savePaths(trainCSV, X_train_paths, y_train)
savePaths(validCSV, X_valid_paths, y_valid)

train_csv = 'csvs/train_set.csv'
valid_csv = 'csvs/valid_set.csv'

X_train_paths, y_train = getSplitData(train_csv)
X_valid_paths, y_valid = getSplitData(valid_csv)
n_classes = max(y_train) + 1
```

数据已被分为训练集和验证集，相应的路径储存在 `X_train_paths` 和 `X_valid_paths` 中，如上述代码的输出所示。

5. 利用独热编码将分类变量表示为二进制向量。独热编码将数据样本的标签表示为具有单个 1 值的零向量，1 值的索引对应于表示有效的真值标签。独热编码是真实标签的一种有用的表示方式，且易用于神经网络，有利于计算模型损失并反向传播。

```
Y_train = oneHotEncode(y_train, n_classes)
y_valid = oneHotEncode(y_valid, n_classes)
```

6. 用以下代码段实现独热编码函数：

```
# one hot encoding
def oneHotEncode(labels, n_classes):
    one_hots = []
    for label in labels:
        one_hot = [0]*n_classes
        if label >= len(one_hot):
            print("Labels out of bounds\nCheck your n_classes parameter")
            return
        one_hot[label] = 1
        one_hots.append(one_hot)
    return np.array(one_hots,dtype=np.float32)
```

7. 由于图像太多而导致无法将所有图像作为 NumPy 数组读入内存，因此创建生成器来分批读取图像到内存中，随机添加的样板能有效地增加数据集：

```
batch_size = 110
add_random_augmentations = False
resize_dims = None
n_train_samples = len(X_train_paths)
train_steps_per_epoch = getSteps(n_train_samples, batch_size, n_augs=1)
n_valid_samples = len(X_valid_paths)
valid_steps_per_epoch = getSteps(n_valid_samples, batch_size, n_augs=0)
train_generator = image_generator(X_train_paths, y_train, batch_size,
                                    resize_dims=resize_dims,
randomly_augment=add_random_augmentations)
valid_generator = image_generator(X_valid_paths, y_valid, batch_size,
                                    resize_dims=resize_dims,
rand_order=False)
```

建模

下图描述了输入经过神经网络的多个层（包括**卷积层**、**子采样层**和**全连接层**），然后最终得出结果的过程：

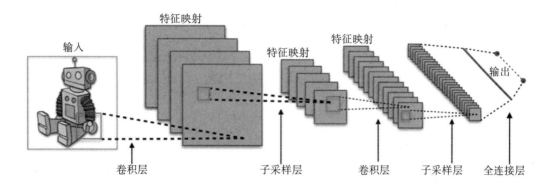

我们创建具有两个全连接层的模型，且第二个全连接层比第一个小，然后是输出层。正如 convModel 所描述的那样，我们在第一层并行运行 3×3、4×4 和 5×5 的过滤器，而在接下来的卷积层中，由于内存限制而减小深度，只运行 3×3 和 5×5 的过滤器。

在 Keras 可以通过逐个添加层帮助我们实现层堆叠，而且在 convModel 中，每层的深度会依次递减。具体是将输入层、批量归一化层、卷积层、最大池化层、丢弃层、**指数线性单位（ELU）层**和 softmax 层堆叠在一起。其中在最大池化层后添加丢弃层是很有必要的，一方面是因为丢弃会对结果产生很大的影响，另一方面是因为如果在最大池化层之前添加，丢弃层的效果可能会减弱。

> 选择 ELU 激活函数是因为它们可以防止神经元"死掉"。

1. 首先，创建 CNN 模型：

```
def convModel(first_conv_shapes=[(4,4),(3,3),(5,5)],
conv_shapes=[(3,3),(5,5)], conv_depths=[12,12,11,8,8],
dense_shapes=[100,50,3], image_shape=(256,256,3), n_labels=3):
    stacks = []
    pooling_filter = (2,2)
    pooling_stride = (2,2)
    inputs = Input(shape=image_shape)
    zen_layer = BatchNormalization()(inputs)
    
    for shape in first_conv_shapes:
        stacks.append(Conv2D(conv_depths[0], shape, padding='same', activation='elu')(zen_layer))
    layer = concatenate(stacks,axis=-1)
    layer = BatchNormalization()(layer)
    layer = MaxPooling2D(pooling_filter,strides=pooling_stride,padding='same')(layer)
```

```python
        layer = Dropout(0.05)(layer)

    for i in range(1,len(conv_depths)):
        stacks = []
        for shape in conv_shapes:
stacks.append(Conv2D(conv_depths[i],shape,padding='same',activation
='elu')(layer))
        layer = concatenate(stacks,axis=-1)
        layer = BatchNormalization()(layer)
        layer = Dropout(i*10**-2+.05)(layer)
        layer = MaxPooling2D(pooling_filter,strides=pooling_stride,
padding='same')(layer)

    layer = Flatten()(layer)
    fclayer = Dropout(0.1)(layer)

    for i in range(len(dense_shapes)-1):
        fclayer = Dense(dense_shapes[i], activation='elu')(fclayer)
        fclayer = BatchNormalization()(fclayer)

    outs = Dense(dense_shapes[-1], activation='softmax')(fclayer)

    return inputs, outs
```

2. 使用前面的 `convModel` 拟合训练集。为了使目标函数最小化,我们将**自适应矩估计(Adam)**优化器作为优化算法。Adam 优化器是 RMSProp 和矩估计的结合,不仅对内存的要求比较低,而且几乎不需要对超参数进行调整。

3. 保持学习率为 `0.001`。如果将学习率设置得很低,也就意味着对权重的更新非常慢,训练的进展也将会非常缓慢。反之,当学习率维持在很高的水平,将会导致损失函数过早收敛于局部值。

4. 使用 `categorical_crossentropy` 损失函数,测量分类模型的性能和准确率,并将结果作为衡量模型性能的指标:

```python
'''
modeling
'''
n_classes = 3
image_shape = (256, 256, 3)

first_conv_shapes = [(4, 4), (3, 3), (5, 5)]
conv_shapes = [(3, 3), (5, 5)]
conv_depths = [12, 12, 11, 8, 8]
dense_shapes = [100, 50, n_classes]

inputs, outs = convModel(first_conv_shapes, conv_shapes,
conv_depths, dense_shapes, image_shape, n_classes)
```

```
model = Model(inputs=inputs, outputs=outs)
learning_rate = .0001
for i in range(20):
    if i > 4:
        learning_rate = .00001  # Anneals the learning rate
    adam_opt = optimizers.Adam(lr=learning_rate)
    model.compile(loss='categorical_crossentropy',
optimizer=adam_opt, metrics=['accuracy'])
    history = model.fit_generator(train_generator,
train_steps_per_epoch, epochs=1,
                        validation_data=valid_generator,
validation_steps=valid_steps_per_epoch, max_queue_size=1)
    model.save('/deeplearning-keras/ch05/weights/model.h5')
```

基于 TensorFlow 后端的输出结果如下所示：

```
Train size:
1744
Valid size:
437
2018-09-10 06:58:45.775834: I
tensorflow/core/platform/cpu_feature_guard.cc:141] Your CPU
supports instructions that this TensorFlow binary was not compiled
to use: SSE4.1 SSE4.2 AVX AVX2 FMA
Epoch 1/1
1/32 [.........................] - ETA: 1:16:46 - loss: 2.0438
-
acc: 0.2545
2/32 [>........................] - ETA: 1:11:16 - loss: 1.9240
- acc: 0.2727
3/32 [=>.......................] - ETA: 1:08:00 - loss: 1.7749
- acc: 0.3091
4/32 [==>......................] - ETA: 1:04:49 - loss: 1.7584
- acc: 0.2909
5/32 [===>.....................] - ETA: 1:01:39 - loss: 1.7117
-acc: 0.3109
6/32 [====>....................] - ETA: 59:20 - loss: 1.6633 -
acc: 0.3303
7/32 [=====>...................] - ETA: 56:58 - loss: 1.6607 -
acc: 0.3221
8/32 [======>..................] - ETA: 54:33 - loss: 1.6422 -
acc: 0.3239
9/32 [=======>.................] - ETA: 52:12 - loss: 1.6201 -
acc: 0.3222
10/32 [========>................] - ETA: 49:50 - loss: 1.6187
- acc: 0.3245
11/32 [=========>...............] - ETA: 47:38 - loss: 1.6232
- acc: 0.3223
12/32 [==========>..............] - ETA: 45:13 - loss: 1.6029
- acc: 0.3242
13/32 [===========>.............] - ETA: 42:55 - loss: 1.5900
- acc: 0.3224
```

```
14/32 [============>.................] - ETA: 40:39 - loss: 1.5931
 - acc: 0.3234
15/32 [=============>................] - ETA: 38:19 - loss: 1.5896
 - acc: 0.3261
16/32 [==============>...............] - ETA: 35:41 - loss: 1.5790
 - acc: 0.3309
17/32 [===============>..............] - ETA: 33:25 - loss: 1.5626
 - acc: 0.3345
18/32 [================>.............] - ETA: 31:10 - loss: 1.5478
 - acc: 0.3366
19/32 [=================>............] - ETA: 28:58 - loss: 1.5372
 - acc: 0.3356
20/32 [==================>...........] - ETA: 26:45 - loss: 1.5302
 - acc: 0.3366


1/32 [..............................] - ETA: 1:11:06 - loss: 0.8727
 - acc: 0.6182
2/32 [>.............................] - ETA: 1:06:38 - loss: 0.8982
 - acc: 0.6091
3/32 [=>............................] - ETA: 1:00:23 - loss: 0.8490
 - acc: 0.6259
4/32 [==>...........................] - ETA: 58:48 - loss: 0.8642 -
acc: 0.6172
5/32 [===>..........................] - ETA: 56:59 - loss: 0.8392 -
acc: 0.6210
6/32 [====>.........................] - ETA: 54:59 - loss: 0.8635 -
acc: 0.6008
7/32 [=====>........................] - ETA: 52:57 - loss: 0.8610 -
acc: 0.6020
8/32 [======>.......................] - ETA: 50:59 - loss: 0.8534 -
acc: 0.5972
```
3/32 [=>............................] - ETA: 57:34 - loss: 0.8460 - acc: 0.6289

该模型在验证集上的准确率约为 62%。

预测

最后，对测试数据集进行预测。首先，加载模型并读入测试图像的路径。然后，使用 ThreadPool 创建一个单独的进程来读入测试图像并调整其大小，从而实现模型在评估样本的同时对图像进行处理。预测结果存储在预测文件（CSV 格式）中以待评估。

```
'''
get predictions
'''
data_path = '/deeplearning-keras/ch05/data/test'
model_path = '/deeplearning-keras/ch05/weights/model.h5'

resize_dims = (256, 256, 3)
test_divisions = 20   # Used for segmenting image evaluation in threading
```

```python
batch_size = 100  # Batch size used for keras predict function

ins, outs = convModel()
model = Model(inputs=ins, outputs=outs)
model.load_weights(model_path)
test_paths, test_labels, _ = readFilePaths(data_path, no_labels=True)
print(str(len(test_paths)) + ' testing images')

pool = ThreadPool(processes=1)
portion = len(test_paths) // test_divisions + 1  # Number of images to read in per pool

async_result = pool.apply_async(convertImages, (test_paths[0 * portion:portion * (0 + 1)],
                                                test_labels[0 * portion:portion * (0 + 1)], resize_dims))
total_base_time = time.time()
test_imgs = []

predictions = []
for i in range(1, test_divisions + 1):
    base_time = time.time()

    print("Begin set " + str(i))
    while len(test_imgs) == 0:
        test_imgs, _ = async_result.get()
    img_holder = test_imgs
    test_imgs = []

    if i < test_divisions:
        async_result = pool.apply_async(convertImages, (test_paths[i * portion:portion * (i + 1)],
                                                        test_labels[0 * portion:portion * (0 + 1)],
                                                        resize_dims))

    predictions.append(model.predict(img_holder, batch_size=batch_size, verbose=0))
    print("Execution Time: " + str((time.time() - base_time) / 60) + 'min\n')

predictions = np.concatenate(predictions, axis=0)
print("Total Execution Time: " + str((time.time() - total_base_time) / 60) + ' mins')

conf = .95  # Prediction confidence
savePredictions = '/deeplearning-keras/ch05/predictions.csv'
predictions = confid(predictions, conf)
header = 'image_name,Type_1,Type_2,Type_3'
save(savePredictions, test_labels, predictions, header)
```

输出如下：

```
image_name,Type_1,Type_2,Type_3
63.jpg, 0.025,0.025,0.95
189.jpg, 0.025,0.95,0.025
77.jpg, 0.025,0.025,0.95
162.jpg, 0.025,0.025,0.95
176.jpg, 0.025,0.025,0.95
88.jpg, 0.025,0.95,0.025
348.jpg, 0.025,0.025,0.95
```

5.3 数字识别

数字识别 MNIST 数据集是由 Yann LeCun、Corinna Cortes 和 Christopher Burges 共同开发的，用于评估机器学习模型在识别手写数字问题上的性能。数字图像取自扫描文件，尺寸已归一化并保证图像居中。每张图片均为 28×28=784 个像素。每个像素都有与其相关联的单个像素值，用来指示该像素的明暗度，值为 0~255 之间（包括 0 和 255），数字越大图像越暗。我们开发数字识别模型流水线，将 10 个数（0~9）作为 10 个类。

5.3.1 准备工作

本小节中，我们将开发模型流水线基于高精度图像识别手写数字（0~9）。模型流水线使用以用于图像分类的 Keras 函数 API 编写的 CNN 模型。

Keras 库提供了一种加载 MNIST 数据的简单方法。数据集以 `mnist.pkl.gz`（15 MB）文件形式自动下载到用户的目录中：

```
from keras.datasets import mnist
# get dataset
(XTrain, yTrain), (XTest, yTest) = mnist.load_data()
```

下载和加载 MNIST 数据集就是调用 `mnist.load_data()` 函数这么简单。

```
# plot 4 images as gray scale
plt.subplot(221)
plt.imshow(XTrain[1], cmap=plt.get_cmap('gray'))
plt.subplot(222)
plt.imshow(XTrain[2], cmap=plt.get_cmap('gray'))
plt.subplot(223)
plt.imshow(XTrain[3], cmap=plt.get_cmap('gray'))
plt.subplot(224)
plt.imshow(XTrain[4], cmap=plt.get_cmap('gray'))
# show the plot
plt.show()
```

运行以上代码得到以下数字图形：

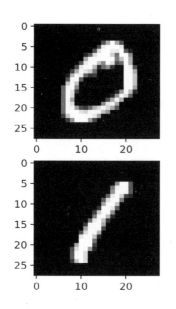
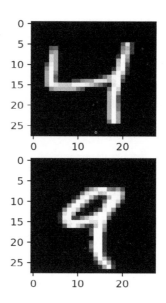

5.3.2 怎么做

让我们为 MNIST 数据集创建一个简单的 CNN，并解释如何使用包括卷积层、池化层和丢弃层等 CNN 实现中的各个部分。随着网络的加深，CNN 会逐渐减少层的维数，并增加特征映射的数量，从而能够在检测更多特征的同时降低计算成本。

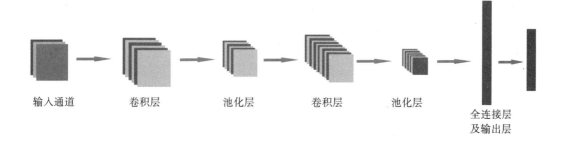

导入所需的 API：

```
import numpy
from keras import backend as K
from keras.utils import np_utils
```

```python
from keras.layers import Dense, Flatten, Dropout
from keras.layers.convolutional import Conv2D, MaxPooling2D
from keras.models import Sequential
import matplotlib.pyplot as plt
from keras.datasets import mnist
from keras.layers.core import Activation
```

建模

1. 首先，初始化随机数生成器为恒定的种子值，以使结果具有可重复性：

```python
numpy.random.seed(0)
```

2. 对 Keras 序贯层进行堆叠处理，以便每个层的输出传输到下一层的过程中不需定义其他数据。导入 `Sequential`：

```python
# create sequential model
model = Sequential()
```

3. 为了提高算法的效率和收敛性，基于最大像素 255 对数据进行归一化处理，将所有像素除以 255 以将输入变为 0～1 之间的数值：

```python
# normalize the dataset
XTrain = XTrain / 255
XTest = XTest / 255
```

4. 根据数据选择合适的算法并得到所需的准确率指标。如果数据与类别的分布比较平衡，可以简单地使用准确率。但如果数据的分布不平衡，那么将无法使用准确率，因为其结果会产生误导，这时就应该使用另一种指标。

```python
# data exploration
print("Number of training examples = %i" % XTrain.shape[0])
print("Number of classes = %i" % len(numpy.unique(yTrain)))
print("Dimension of images = {:d} x {:d} ".format(XTrain[1].shape[0], XTrain[1].shape[1]))
unique, count = numpy.unique(yTrain, return_counts=True)
print("The number of occurrences of each class in the dataset = %s " % dict(zip(unique, count)), "\n")
```

5. 输出为：

```
Number of training examples = 60000
Number of classes = 10
Dimension of images = 28 x 28
The number of occurrences of each class in the dataset = {0: 5923,
1: 6742, 2: 5958, 3: 6131, 4: 5842, 5: 5421, 6: 5918, 7: 6265, 8:
5851, 9: 5949}
```

我们可以看到数据集由 60 000 个训练样本组成，其中每个样本都是 28×28 的图像。由于类的分布是平衡的，我们可以使用准确率作为度量标准。

6. 重塑数据样本或图像，以使它们适合用卷积神经网络进行训练。在 keras 库中，各层使用图像像素的维度为 (*pixels*)(*width*)(*height*)。而在 MNIST 中，像素值就是灰度值，所以像素的维度为 1。

7. 以独热编码的形式输出结果，这意味着将有从 0 ～ 9 共 10 个类，每个数字为一个类：

```
XTrain = XTrain.reshape(XTrain.shape[0], 28, 28,
1).astype('float32')
XTest = XTest.reshape(XTest.shape[0], 28, 28, 1).astype('float32')
yTrain = np_utils.to_categorical(yTrain)
yTest = np_utils.to_categorical(yTest)
```

8. 用最简单的架构实现 CNN 的第一层。对于序贯模型，对层进行堆叠，并在第一层，也就是卷积层 Conv2D() 中指定图像的输入维度、特征映射数量、输入形状和激活函数（rule），然后添加内核维度为 2×2 的最大池化层：

```
# modeling
model.add(Conv2D(40, kernel_size=5, padding="same",
input_shape=(28, 28, 1), activation='relu'))
model.add(Conv2D(50, kernel_size=5, padding="valid",
activation='relu'))
model.add(MaxPooling2D(pool_size=(2, 2), strides=(2, 2)))
model.add(Dropout(0.2))
```

9. 添加一个平展层，用于接收 CNN 的输出并将其平展，再将平展后的数据作为稠密层的输入，进而将其传递到输出层。在输出层使用 softmax 输出多类分类的预测概率向量：

```
model.add(Flatten())
model.add(Dense(100))
model.add(Activation("relu"))
model.add(Dropout(0.2))
model.add(Dense(10))
model.add(Activation("softmax"))
```

10. 最后，编译模型并使用 fit() 进行训练，拟合训练数据和类别，设置迭代次数和批处理大小（每个训练周期的图像数）。作为构建模型的最后一步，我们需要对模型进行评估以确保不会出现数据过拟合的情况。模型评估指的是使用训练过程中得到的权重来预测测试数据集的值进而确定模型在未知数据上的性能。我们使用 `categorical_crossentropy` 作为该模型的成本函数，如以下代码所示：

```
model.compile(loss='categorical_crossentropy', optimizer='adam',
metrics=['accuracy'])
model.fit(XTrain, yTrain, epochs=32, batch_size=200,
validation_split=0.2)
scores = model.evaluate(XTest, yTest, verbose=10)
print(scores)
```

11. 以下是通过训练 48 000 个样本并在 12 000 个样本上进行验证获得的输出：

```
200/48000 [..............................] - ETA: 33:31 - loss: 2.3074 - acc: 0.0650
400/48000 [..............................] - ETA: 32:15 - loss: 2.2640 - acc: 0.1350
600/48000 [..............................] - ETA: 32:12 - loss: 2.2285 - acc: 0.1600
800/48000 [..............................] - ETA: 32:40 - loss: 2.1714 - acc: 0.1975
1000/48000 [..............................] - ETA: 33:09 - loss: 2.0927 - acc: 0.2650
48000/48000 [==============================] - 2382s 50ms/step - loss: 0.2471 - acc: 0.9238
...
...
47600/48000 [============================>.] - ETA: 17s - loss: 0.0067 - acc: 0.9978
47800/48000 [============================>.] - ETA: 8s - loss: 0.0068 - acc: 0.9977
48000/48000 [==============================] - 2237s 47ms/step - loss: 0.0069 - acc: 0.9977
```

我们可以在 CPU 上运行这个神经网络，并获得上述输出。本网络的架构虽然相对简单，但准确率可以达到 99% 以上。

CHAPTER 6

第 6 章

生成式对抗网络

本章包括以下内容：
- 基础生成式对抗网络
- 边界搜索生成式对抗网络
- 深度卷积生成式对抗网络

6.1 引言

生成式对抗网络（GAN）是深度学习的最新发展之一。2014 年，Ian Goodfellow 首次提出了 GAN（https://arxiv.org/pdf/1406.2661.pdf），通过同时训练两个深度网络（**生成器**和**判别器**）来解决无监督学习的问题。在整个训练过程中，两个网络之间相互进行竞争和合作，最终能够更精准地完成任务。

GAN 中的两个网络所扮演的角色就像是造假者（生成器）和警察（判别器）。最初，造假者向警察出示假钱，警察判定这是假的，并向造假者提供反馈意见，说明这钱为什么是假的。然后造假者根据收到的反馈信息再制作新的假钱，警察再次判定这笔钱是假的，并提供更多的反馈。根据最新的反馈，造假者再次尝试。这个循环无限地持续下去，直到最后钱真实到警察也难以判断真假的程度。

本章将介绍有关 GAN 的详细内容。

GAN 概述

对抗模型框架可在多层感知器模型上直接应用。为了在数据 x 上学习生成器的分布 p_g，在输入噪声变量上定义先验分布 $p(z)$，然后将数据空间的映射表示为 $G(z, \theta_g)$，其中 G 是一个可微函数，可由多层感知器表示，参数为 θ_g。定义第二个多层感知器，称为判别器：$D(x, \theta_g)$，输出为单标量。$D(x)$ 表示 x 来自真实数据的概率而不是 p_g。目标是训练 D，使训练样本和来自 G 的样本进行正确分类的概率最大化，同时训练 G 来最小化 $\log(1-D(G(Z))$：

$$\min_G \max_D V(D, G) = \mathbb{E}_{x \sim p_{data}(x)}[\log D(x)] + \mathbb{E}_{z \sim p_z(z)}[\log(1 - D(G(z)))]$$

6.2 基本的生成式对抗网络

本节将介绍用于 Fashion-MNIST 数据集的基础 GAN。

Fashion-MNIST 是一个 Zalando 物品图像数据集，包含 60 000 个样本组成的训练集以及 10 000 个样本组成的测试集。每个样本都是一个 28×28 的灰度图像，分别与 10 个类别的标签关联。

以下是 Fashion-MNIST 的一些示例图像：

Fashion-MNIST 在 Keras 中直接可用。

6.2.1 准备工作

创建一个名为 GAN 的类，导入相关的类并初始化变量：

```
from __future__ import print_function, division
from keras.datasets import fashion_mnist
from keras.layers import Input, Dense, Reshape, Flatten, Dropout
from keras.layers import BatchNormalization, Activation, ZeroPadding2D
from keras.layers.advanced_activations import LeakyReLU
from keras.layers.convolutional import UpSampling2D, Conv2D
from keras.models import Sequential, Model
from keras.optimizers import Adam
import matplotlib.pyplot as plt
import sys
import numpy as np

GAN class():
....
```

定义要在 __init__() 中使用的常量：

```
self.img_rows = 28
 self.img_cols = 28
self.channels = 1
self.img_shape = (self.img_rows, self.img_cols, self.channels)
self.latent_dim = 100
```

其中 `img_shape` 为 (28,28,1)，`latent_dim` 为 100。

6.2.2 怎么做

以下部分介绍了 Fashion-MNIST 数据集上的 GAN。

构建生成器

创建包含以下层的序贯模型：

- 输入为 (`self.latent_dim`)、输出为 (*，256 个单位) 的密集层。
- 用于应用此函数到输入数据的 Leaky ReLU 层。
- 批量归一化层：归一化数据。
- 512 密集层：输出为 (*，512 个单元) 的层。
- 批量归一化层。

- 密集层（*，1024）。
- Leaky ReLU 层。
- 批量归一化层。
- 密集层，大小为（*，256），使用的激活函数为 tanh。
- 形变为 `img_shape` 的层。
- 在模型中添加噪声的层，shape = (self.latent_dim,)。

```
def build_generator(self):
model = Sequential()
model.add(Dense(256, input_dim=self.latent_dim))
model.add(LeakyReLU(alpha=0.2))
model.add(BatchNormalization(momentum=0.8))
model.add(Dense(512))
model.add(LeakyReLU(alpha=0.2))
model.add(BatchNormalization(momentum=0.8))
model.add(Dense(1024))
model.add(LeakyReLU(alpha=0.2))
model.add(BatchNormalization(momentum=0.8))
model.add(Dense(np.prod(self.img_shape), activation='tanh'))
model.add(Reshape(self.img_shape))
model.summary()
noise = Input(shape=(self.latent_dim,))
img = model(noise)
return Model(noise, img)
```

下图展示了噪声如何在生成器中被转换成图像：

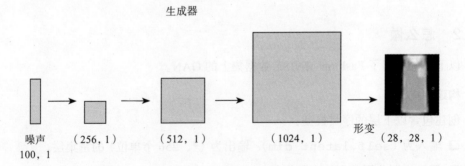

以下代码列出了模型总结：

```
Layer (type) Output Shape Param #
=================================================================
dense_4 (Dense) (None, 256) 25856
```

```
leaky_re_lu_3 (LeakyReLU)    (None, 256)         0
_____
batch_normalization_1 (Batch (None, 256)         1024
_____
dense_5 (Dense)              (None, 512)         131584
_____
leaky_re_lu_4 (LeakyReLU)    (None, 512)         0
_____
batch_normalization_2 (Batch (None, 512)         2048
_____
dense_6 (Dense)              (None, 1024)        525312
_____
leaky_re_lu_5 (LeakyReLU)    (None, 1024)        0
_____
batch_normalization_3 (Batch (None, 1024)        4096
_____
dense_7 (Dense)              (None, 784)         803600
_____
reshape_1 (Reshape)          (None, 28, 28, 1)   0
=================================================================
Total params: 1,493,520
Trainable params: 1,489,936
Non-trainable params: 3,584
```

构建判别器

可以通过相反的顺序使用序贯模型构建判别器：

❑ 第一层将 `input_shape` 平展为（28，28，1）。

❑ 添加输出为（*，512）的密集层。

❑ 添加 Leaky ReLU 激活函数。

❑ 添加另一个密集层，输出为（*，256）。

❑ 添加另一个激活函数 Leaky ReLU。

❑ 添加形状为（*，1）的最终输出。

```python
def build_discriminator(self):
    model = Sequential()
    model.add(Flatten(input_shape=self.img_shape))
    model.add(Dense(512))
    model.add(LeakyReLU(alpha=0.2))
    model.add(Dense(256))
    model.add(LeakyReLU(alpha=0.2))
    model.add(Dense(1, activation='sigmoid'))
    model.summary()
    img = Input(shape=self.img_shape)
    validity = model(img)
    return Model(img, validity)
```

初始化 GAN 实例

自定义一个 GAN 类，用生成器和判别器进行初始化，具体步骤如下：

1. 初始化变量 `img_rows`、`img_cols`、`channels`、`img_shape` 和 `latent_dim`。
2. 初始化优化器，本例使用 Adam 优化器。
3. 实例化判别器：
   - 使用 `build_discriminator()` 进行实例化。
   - 判别器，其使用的损失函数为 `binary_crossentropy`，优化器为 Adam，指标为准确率。
4. 生成器：
   - 使用 `build_generator()` 进行实例化。
   - 将含噪声的生成图像作为输入。
5. 利用判别器检查图像的真伪。
6. 组合模型：利用生成器欺骗判别器。
   - 生成器的输入为 z，输入尺寸为 `(*,self.latent_dim)`。
   - 生成器的输出作为判别器的输入。
7. 编译组合模型得到其损失。

```
def __init__(self):
    self.img_rows = 28
    self.img_cols = 28
    self.channels = 1
    self.img_shape = (self.img_rows, self.img_cols, self.channels)
    self.latent_dim = 100
    optimizer = Adam(0.0002, 0.5)
    # Build and compile the discriminator
    self.discriminator = self.build_discriminator()
    self.discriminator.compile(loss='binary_crossentropy',
        optimizer=optimizer,
        metrics=['accuracy'])
    # Build the generator
    self.generator = self.build_generator()
    # The generator takes noise as input and generates imgs
    z = Input(shape=(self.latent_dim,))
    img = self.generator(z)
    # For the combined model we will only train the generator
    self.discriminator.trainable = False
    # The discriminator takes generated images as input and
determines validity
    validity = self.discriminator(img)
```

```
# The combined model (stacked generator and discriminator)
# Trains the generator to fool the discriminator
self.combined = Model(z, validity)
self.combined.compile(loss='binary_crossentropy',
optimizer=optimizer)
```

接下来我们将讨论如何将上述内容组合在一起,并迭代训练GAN,编译得到其损失并存储生成的图像。

训练GAN

模型的训练过程:

1. 首先,加载数据集。

2. 对数据进行缩放。

3. 生成结果(为真或假)。

4. 每次迭代过程中执行以下操作。

  ❑ 随机选择一批图像。

  ❑ 生成图像。

  ❑ 计算真实图像和虚假图像的损失。

  ❑ 采样和绘图。

```
def train(self, epochs, batch_size=128, sample_interval=50):

    # Load the dataset
    (X_train, _), (_, _) = fashion_mnist.load_data()

    # Rescale -1 to 1
    X_train = X_train / 127.5 - 1.
    X_train = np.expand_dims(X_train, axis=3)

    # Adversarial ground truths
    valid = np.ones((batch_size, 1))
    fake = np.zeros((batch_size, 1))

    for epoch in range(epochs):

        # ---------------------
        #  Train Discriminator
        # ---------------------

        # Select a random batch of images
        idx = np.random.randint(0, X_train.shape[0], batch_size)
        imgs = X_train[idx]
```

```
        noise = np.random.normal(0, 1, (batch_size, self.latent_dim))

        # Generate a batch of new images
        gen_imgs = self.generator.predict(noise)

        # Train the discriminator
        d_loss_real = self.discriminator.train_on_batch(imgs, valid)
        d_loss_fake = self.discriminator.train_on_batch(gen_imgs, fake)
        d_loss = 0.5 * np.add(d_loss_real, d_loss_fake)

        # ---------------------
        #  Train Generator
        # ---------------------
        noise = np.random.normal(0, 1, (batch_size, self.latent_dim))

        # Train the generator (to have the discriminator label samples as valid)
        g_loss = self.combined.train_on_batch(noise, valid)

        # Plot the progress
        print ("%d [D loss: %f, acc.: %.2f%%] [G loss: %f]" % (epoch, d_loss[0], 100*d_loss[1], g_loss))

        # If at save interval => save generated image samples
        if epoch % sample_interval == 0:
            self.sample_images(epoch)
```

输出绘图

第一次迭代后的图像：

经过 9 000 次迭代后的图像：

经过 29 800 次迭代后的图像：

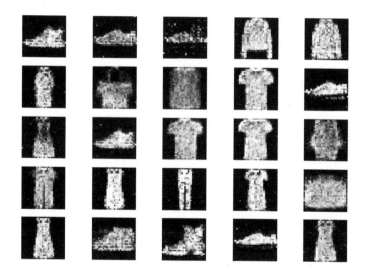

GAN 的平均指标

模型经过 30 000 次迭代后，可以获得如下指标：

```
mean d loss:0.6404680597275333
mean g loss:0.9513815407413333
mean d accuracy:62.71046875
```

判别器的平均损失约为 0.64，生成器平均损失为 0.95，判别器准确率约为 62%。迭代过程中每一次的指标如图所示：

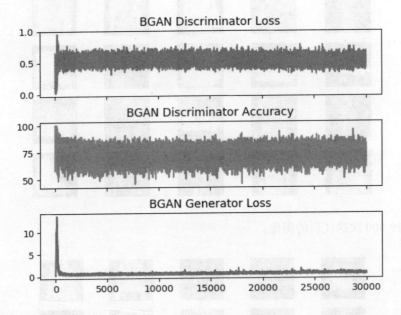

接下来我们介绍一个更高级的 GAN，称为**边界搜索生成式对抗网络（BGAN）**。

6.3 边界搜索生成式对抗网络

在最初的 GAN 网络的论文中，目标函数为：

$$\min_G \max_D V(D, G) = \mathbb{E}_{x \sim p_{data}(x)}[\log D(x)] + \mathbb{E}_{z \sim p_z(z)}[\log(1 - D(G(z)))]$$

其中：

- x：数据
- p_g：数据 x 上的生成器的分布
- $p(z)$：输入噪声变量的先验分布
- $G(z, \theta_g)$：先验分布到数据空间的映射
- G：用多层感知器和参数 θ_g 表示的可微函数

- $D(x, \theta_d)$：判别器，输出单标量的第二层多层感知器
- $D(x)$：表示 x 来自数据的概率而不是 p_g

目标是训练 D 以使 G 中样本和训练样本获得正确标签的概率最大。同时，通过训练 G 以最小化 $\log(1-D(G(Z)))$；最优判别器 $D_G^*(x)$ 表示如下：

$$D_G^*(x) = \frac{p_{data}(x)}{p_{data}(x) + p_g(x)}$$

其中 $p_{data}(x)$ 是实际的分布，可通过重排样本中的项得到：

$$p_{data}(x) = p_g(x)\frac{D_G^*(x)}{1-D_G^*(x)}$$

假设训练 $D(x)$ 的次数越多，它就越接近 $D_G^*(x)$，并且 GAN 的训练也会更优。对于生成器而言，当 $p_{data}(x) = p_g(x)$ 时，取得最优值 0.5，$D(x) = 0.5$ 是决策边界。我们希望生成 $x \sim G(x)$ 以接近决策边界。因此，该方法被称为边界搜索 GAN。

如果令 $D(x) = 1-D(x) = 0.5$，则目标函数如下所示：

$$\min_G \mathbb{E}_{z \sim p_z(z)}\left[\frac{1}{2}\left(\log D(x) - \log(1-D(x))\right)^2\right]$$

求 $D(x)$ 的对数，因为此时 $D(x)$ 为概率分布。在本节中，我们将探讨如何在 Keras 中实现边界搜索 GAN。

6.3.1 准备工作

创建一个名为 BGAN 的类，导入相关类并初始化变量：

```
from __future__ import print_function, division

from keras.datasets import fashion_mnist
from keras.layers import Input, Dense, Reshape, Flatten, Dropout
from keras.layers import BatchNormalization, Activation, ZeroPadding2D
from keras.layers.advanced_activations import LeakyReLU
from keras.layers.convolutional import UpSampling2D, Conv2D
from keras.models import Sequential, Model
from keras.optimizers import Adam
import keras.backend as K
import matplotlib.pyplot as plt
import sys
import numpy as np
```

```
class BGAN():
....
```

如下所示,定义所用常量:

```
self.img_rows = 28
self.img_cols = 28
self.channels = 1
self.img_shape = (self.img_rows, self.img_cols, self.channels)
self.latent_dim = 100
```

其中 `img_shape` 为 `(28,28,1)`,`latent_dim` 为 `100`。

6.3.2 怎么做

创建生成器和判别器,比较损失值,进行迭代。

生成器

创建一个包含以下层的序贯模型:

- 一个输入为(`self.latent_dim`),输出为(*,256 个单元)的密集层。
- 用于应用此函数到输入数据的 Leaky ReLU 层。
- 批量归一化层:归一化数据。
- 512 密集层:输出为(*,512 个单元)的层。
- 用于应用此函数到输入数据的 Leaky ReLU 层。
- 批量归一化层。
- 密集层(*,1024)。
- Leaky ReLU。
- 批量归一化层。
- 密集层:大小为(*,256),使用的激活函数为 `tanh`。
- 形变为 img_shape 的层。
- 在模型中添加噪声的层,`shape=(self.latent_dim,)`。

```
def build_generator(self):
    model = Sequential()
    model.add(Dense(256, input_dim=self.latent_dim))
    model.add(LeakyReLU(alpha=0.2))
    model.add(BatchNormalization(momentum=0.8))
    model.add(Dense(512))
```

```
model.add(LeakyReLU(alpha=0.2))
model.add(BatchNormalization(momentum=0.8))
model.add(Dense(1024))
model.add(LeakyReLU(alpha=0.2))
model.add(BatchNormalization(momentum=0.8))
model.add(Dense(np.prod(self.img_shape), activation='tanh'))
model.add(Reshape(self.img_shape))
model.summary()
noise = Input(shape=(self.latent_dim,))
img = model(noise)
return Model(noise, img)
```

接下来介绍判别器,它将用于检验生成的图像与真实图像的近似程度。

判别器

判别器是一个反方向的序贯模型:

- 第一层将 `input_shape` 平展为(28,28,1)。
- 添加一个输出为(*,512)的密集层。
- 添加 Leaky ReLU 激活函数。
- 添加另一个输出为(*,256)的密集层。
- 再添加一个 Leaky ReLU 激活函数。
- 添加形状为(*,1)且包含 `sigmoid` 激活函数的最终输出。

```
def build_discriminator(self):
    model = Sequential()
    model.add(Flatten(input_shape=self.img_shape))
    model.add(Dense(512))
    model.add(LeakyReLU(alpha=0.2))
    model.add(Dense(256))
    model.add(LeakyReLU(alpha=0.2))
    model.add(Dense(1, activation='sigmoid'))
    model.summary()
    img = Input(shape=self.img_shape)
    validity = model(img)
    return Model(img, validity)
```

接下来我们将介绍如何初始化 BGAN 类。

初始化 BGAN 类

自定义一个 GAN 类,用生成器和判别器进行初始化,具体步骤如下:

1. 初始化变量 `img_rows`、`img_cols`、`channels`、`img_shape` 和 `latent_dim`。
2. 初始化优化器,本例使用 Adam 优化器。

3. 实例化判别器。
   - 使用 `build_discriminator()` 进行实例化。
   - 编译判别器，其使用的损失函数为 `binary_crossentropy`，优化器为 Adam，指标为准确度。
4. 生成器：
   - 使用 `build_generator()` 进行实例化。
   - 将含噪声的生成图像作为输入。
5. 利用判别器检查图像的真伪。
6. 组合模型：利用生成器欺骗判别器。
   - 生成器的输入为 z，输入形状为 `(*,self.latent_dim)`。
   - 生成器的输出作为判别器的输入。
7. 编译组合模型得到其损失。此处开始变得有趣起来，注意使用了边界搜索损失：

```
class BGAN():

    def __init__(self):
        self.img_rows = 28
        self.img_cols = 28
        self.channels = 1
        self.img_shape = (self.img_rows, self.img_cols, self.channels)
        self.latent_dim = 100

        optimizer = Adam(0.0002, 0.5)

        # Build and compile the discriminator
        self.discriminator = self.build_discriminator()
        self.discriminator.compile(loss='binary_crossentropy',
            optimizer=optimizer,
            metrics=['accuracy'])

        # Build the generator
        self.generator = self.build_generator()

        # The generator takes noise as input and generated imgs
        z = Input(shape=(100,))
        img = self.generator(z)

        # For the combined model we will only train the generator
        self.discriminator.trainable = False

        # The valid takes generated images as input and determines validity
```

```
        valid = self.discriminator(img)

        # The combined model (stacked generator and discriminator)
        # Trains the generator to fool the discriminator
        self.combined = Model(z, valid)
        self.combined.compile(loss=self.boundary_loss,
optimizer=optimizer)
```

边界搜索损失

正如本节开头部分所解释的那样，通过以下公式实现边界搜索损失：

$$\min_{G} \mathbb{E}_{z \sim p_z(z)} \left[\frac{1}{2} \left(\log D(x) - \log(1 - D(x)) \right)^2 \right]$$

其中 $D(x)$ 是 x 来自数据而不是 p_g 的概率。

> 关于边界损失搜索更多信息，请参阅以下链接：https//wiseodd.github.io/techblog/2017/03/07/boundary-seeking-gan/。

与上一个示例相比，这是唯一的变化。在构建的类中用如下函数实现：

```
import keras.backend as K

def boundary_loss(self, y_true, y_pred):
 return 0.5 * K.mean((K.log(y_pred) - K.log(1 - y_pred))**2)
```

接下来将介绍如何使用训练集对网络进行训练。

训练 BGAN

模型训练：

1. 首先，加载数据集。

2. 对数据进行缩放。

3. 生成判断结果（真或假）。

4. 每次迭代过程中执行以下操作：

    ❑ 随机选择一批图像

    ❑ 生成图像

    ❑ 计算真实和虚假图像的损失

    ❑ 采样和绘图

```python
def train(self, epochs, batch_size=128, sample_interval=50):

    # Load the dataset
    (X_train, _), (_, _) = fashion_mnist.load_data()

    # Rescale -1 to 1
    X_train = X_train / 127.5 - 1.
    X_train = np.expand_dims(X_train, axis=3)

    # Adversarial ground truths
    valid = np.ones((batch_size, 1))
    fake = np.zeros((batch_size, 1))

    for epoch in range(epochs):

        # ---------------------
        #  Train Discriminator
        # ---------------------

        # Select a random batch of images
        idx = np.random.randint(0, X_train.shape[0], batch_size)
        imgs = X_train[idx]

        noise = np.random.normal(0, 1, (batch_size, self.latent_dim))

        # Generate a batch of new images
        gen_imgs = self.generator.predict(noise)

        # Train the discriminator
        d_loss_real = self.discriminator.train_on_batch(imgs, valid)
        d_loss_fake = self.discriminator.train_on_batch(gen_imgs, fake)
        d_loss = 0.5 * np.add(d_loss_real, d_loss_fake)


        # ---------------------
        #  Train Generator
        # ---------------------

        g_loss = self.combined.train_on_batch(noise, valid)

        # Plot the progress
        print ("%d [D loss: %f, acc.: %.2f%%] [G loss: %f]" % (epoch, d_loss[0], 100*d_loss[1], g_loss))

        # If at save interval => save generated image samples
        if epoch % sample_interval == 0:
            self.sample_images(epoch)
```

输出绘图

迭代到不同次数时的绘图。

迭代 0 次

初次迭代生成的输出图像：

迭代 10 000 次

迭代 10 000 次后生成的图像。值得注意的是，图像比使用简单 GAN 更清晰：

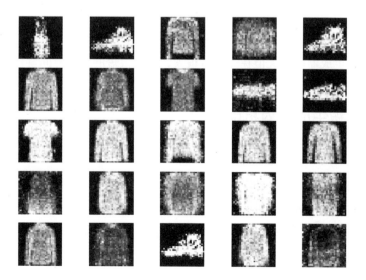

BGAN 模型的指标

计算 D 的平均损失、准确率，G 的平均损失。其中 D 为**判别器**，G 为**生成器**：

```
mean d loss:0.5253478690446334
mean g loss:0.5571205118226668
mean d accuracy:72.77859375
```

可以看出，此时 D 的准确率比简单 GAN 的高很多。

绘制指标

绘制各个平均值指标的图像：

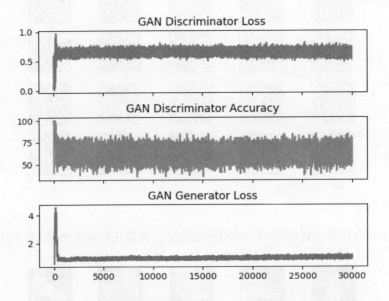

可以看到，生成器的损失值在开始时很大，但在很短的一段时间就稳定了下来。

6.4 深度卷积生成式对抗网络

曾经有一段时间，使用卷积神经网络（CNN）的 GAN 网络并不成功，直到论文作者提出了以下改进方法。

以下是稳定的**深度卷积生成式对抗网络（DCGAN）**的架构：

❏ 判别器中用跨步卷积代替池化层，生成器中用小数步长卷积（又称反卷积）代替池化层

- 在生成器和判别器中使用批量归一化
- 删除全连接隐藏层以获得更深架构
- 对生成器中除输出之外的所有层使用 ReLU 激活函数，输出层使用 tanh 函数
- 对判别器中所有层使用 LeakyReLU 激活函数

同样使用 Fashion-MNIST 数据集进行模型构建。

6.4.1 准备工作

导入相关类并初始化 DCGAN，具体如下所示：

```
from __future__ import print_function, division

from keras.datasets import fashion_mnist
from keras.layers import Input, Dense, Reshape, Flatten, Dropout
from keras.layers import BatchNormalization, Activation, ZeroPadding2D
from keras.layers.advanced_activations import LeakyReLU
from keras.layers.convolutional import UpSampling2D, Conv2D
from keras.models import Sequential, Model
from keras.optimizers import Adam
import matplotlib.pyplot as plt
import sys
import numpy as np

class DCGAN():
    ...
```

定义需要使用的常量：

```
class DCGAN():
    def __init__(self):
        # Input shape
        self.img_rows = 28
        self.img_cols = 28
        self.channels = 1
        self.img_shape = (self.img_rows, self.img_cols, self.channels)
        self.latent_dim = 100
```

以下部分将在后面的章节中探讨：

- 构建生成器
- 构建判别器
- 循环迭代并评估损失

6.4.2 怎么做

我们先来看生成器。

生成器

创建一个包含 Leaky ReLU 激活函数的双卷积层的生成器：

- 输出为（128 * 7 * 7）的密集层
- 将输出形变为（7，7，128）
- 2D 上采样（上采样是指将图像采样到更高分辨率，2D 表示上采样两个二维图像）
- 2D 卷积，输出为 128 维
- 批量归一化
- RELU 激活
- 2D 上采样
- 2D 卷积，输出为 64 维
- 批量归一化
- RELU 激活
- 输出为三维的 2D 卷积
- 最后的 tanh 激活

```
def build_generator(self):

    model = Sequential()
    model.add(Dense(128 * 7 * 7, activation="relu", input_dim=self.latent_dim))
    model.add(Reshape((7, 7, 128)))
    model.add(UpSampling2D())
    model.add(Conv2D(128, kernel_size=3, padding="same"))
    model.add(BatchNormalization(momentum=0.8))
    model.add(Activation("relu"))
    model.add(UpSampling2D())
    model.add(Conv2D(64, kernel_size=3, padding="same"))
    model.add(BatchNormalization(momentum=0.8))
    model.add(Activation("relu"))
    model.add(Conv2D(self.channels, kernel_size=3, padding="same"))
    model.add(Activation("tanh"))
    model.summary()
    noise = Input(shape=(self.latent_dim,))
    img = model(noise)

    return Model(noise, img)
```

生成器总结

先平展输入 633472,输出形状为(28,28,1)的图像:

```
Layer (type) Output Shape Param #
=================================================================
dense_2 (Dense) (None, 6272) 633472

reshape_1 (Reshape) (None, 7, 7, 128) 0

up_sampling2d_1 (UpSampling2 (None, 14, 14, 128) 0

conv2d_5 (Conv2D) (None, 14, 14, 128) 147584

batch_normalization_4 (Batch (None, 14, 14, 128) 512

activation_1 (Activation) (None, 14, 14, 128) 0

up_sampling2d_2 (UpSampling2 (None, 28, 28, 128) 0

conv2d_6 (Conv2D) (None, 28, 28, 64) 73792

batch_normalization_5 (Batch (None, 28, 28, 64) 256

activation_2 (Activation) (None, 28, 28, 64) 0

conv2d_7 (Conv2D) (None, 28, 28, 1) 577

activation_3 (Activation) (None, 28, 28, 1) 0
=================================================================
Total params: 856,193
Trainable params: 855,809
Non-trainable params: 384
```

训练生成器

训练是通过 `__init__(self)` 函数实现的,此处使用 Adam 优化器:

1. 首先,使用 `build_generator()` 构建生成器。

2. 然后,编译 `self.generator`,损失函数为 `binary_crossentropy`,用已定义的优化器,指标为 `accuracy`。

```python
def __init__(self):
    ....
optimizer = Adam(0.0002, 0.5)
# Build the generator
self.generator = self.build_generator()
```

```
# The generator takes noise as input and generates imgs
z = Input(shape=(100,))
img = self.generator(z)
```

接下来介绍判别器。

判别器

我们先来看下判别器是如何构建的。

构建判别器

判别器从图像开始反向进行，最后输出损失值：

```
def build_discriminator(self):
    model = Sequential()
    model.add(Conv2D(32, kernel_size=3, strides=2, input_shape=self.img_shape, padding="same"))
    model.add(LeakyReLU(alpha=0.2))
    model.add(Dropout(0.25))
    model.add(Conv2D(64, kernel_size=3, strides=2, padding="same"))
    model.add(ZeroPadding2D(padding=((0,1),(0,1))))
    model.add(BatchNormalization(momentum=0.8))
    model.add(LeakyReLU(alpha=0.2))
    model.add(Dropout(0.25))
    model.add(Conv2D(128, kernel_size=3, strides=2, padding="same"))
    model.add(BatchNormalization(momentum=0.8))
    model.add(LeakyReLU(alpha=0.2))
    model.add(Dropout(0.25))
    model.add(Conv2D(256, kernel_size=3, strides=1, padding="same"))
    model.add(BatchNormalization(momentum=0.8))
    model.add(LeakyReLU(alpha=0.2))
    model.add(Dropout(0.25))
    model.add(Flatten())
    model.add(Dense(1, activation='sigmoid'))
    model.summary()

    img = Input(shape=self.img_shape)
    validity = model(img)

    return Model(img, validity)
```

判别器总结

运行时，判别器总结如下所示：

```
Layer (type)          Output Shape         Param #
=================================================================
conv2d_1 (Conv2D)     (None, 14, 14, 32)   320
_____
leaky_re_lu_1 (LeakyReLU) (None, 14, 14, 32)   0
_____
```

```
dropout_1 (Dropout)         (None, 14, 14, 32)    0
conv2d_2 (Conv2D)           (None, 7, 7, 64)      18496
zero_padding2d_1 (ZeroPaddin (None, 8, 8, 64)     0
batch_normalization_1 (Batch (None, 8, 8, 64)     256
leaky_re_lu_2 (LeakyReLU)   (None, 8, 8, 64)      0
dropout_2 (Dropout)         (None, 8, 8, 64)      0
conv2d_3 (Conv2D)           (None, 4, 4, 128)     73856
batch_normalization_2 (Batch (None, 4, 4, 128)    512
leaky_re_lu_3 (LeakyReLU)   (None, 4, 4, 128)     0
dropout_3 (Dropout)         (None, 4, 4, 128)     0
conv2d_4 (Conv2D)           (None, 4, 4, 256)     295168
batch_normalization_3 (Batch (None, 4, 4, 256)    1024
leaky_re_lu_4 (LeakyReLU)   (None, 4, 4, 256)     0
dropout_4 (Dropout)         (None, 4, 4, 256)     0
flatten_1 (Flatten)         (None, 4096)          0
dense_1 (Dense)             (None, 1)             4097
=================================================================
Total params: 393,729
Trainable params: 392,833
Non-trainable params: 896
```

编译判别器

类似于生成器,判别器利用 __init__(self) 函数进行编译:

```
def __init__(self):
 ....
 optimizer = Adam(0.0002, 0.5)

 # Build and compile the discriminator
 self.discriminator = self.build_discriminator()
 self.discriminator.compile(loss='binary_crossentropy',
 optimizer=optimizer,
 metrics=['accuracy'])

 ...
 self.discriminator.trainable = False
```

```
    # The discriminator takes generated images as input and determines
validity
    valid = self.discriminator(img)
```

组合模型——生成器和判别器

通过创建一个组合模型(生成器和判别器)来使生成效果更好。以下为 `__init__(self)` 函数:

```
def __init__(self):
....
optimizer = Adam(0.0002, 0.5)

# Build and compile the discriminator
self.discriminator = self.build_discriminator()
self.discriminator.compile(loss='binary_crossentropy',
optimizer=optimizer,
metrics=['accuracy'])

# Build the generator
self.generator = self.build_generator()

# The generator takes noise as input and generates imgs
z = Input(shape=(100,))
img = self.generator(z)

# For the combined model we will only train the generator
    self.discriminator.trainable = False
    # The discriminator takes generated images as input and determines
validity
    valid = self.discriminator(img)

    # The combined model (stacked generator and discriminator)
    # Trains the generator to fool the discriminator
    self.combined = Model(z, valid)
    self.combined.compile(loss='binary_crossentropy', optimizer=optimizer)
```

下图显示了组合模型的结构:

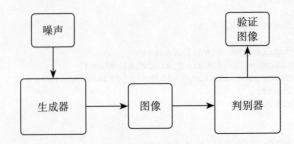

根据判别器的反馈对生成器进行训练

本节将研究如何计算生成器损失,并通过将真实图像和验证图像之间的损失馈入判别器来使其变得更智能:

- 首先,创建对抗式平台:验证图像和伪造图像的数据持有者。
- 迭代:
  - 选择一组随机图像
  - 使用噪声作为生成器的输入,并生成结果图像
  - 将真实图像和验证图像进行比较来计算虚假图像和验证图像之间的 `d_loss`
  - 计算总的判别器损失的平均值
  - 计算堆叠在彼此之上的生成器和判别器的组合损失

```python
def train(self, epochs, batch_size=128, save_interval=50):
    # Load the dataset
    (X_train, _), (_, _) = fashion_mnist.load_data()

    # Rescale -1 to 1
    X_train = X_train / 127.5 - 1.
    X_train = np.expand_dims(X_train, axis=3)

    # Adversarial ground truths
    valid = np.ones((batch_size, 1))
    fake = np.zeros((batch_size, 1))

    for epoch in range(epochs):

        # ---------------------
        #  Train Discriminator
        # ---------------------

        # Select a random half of images
        idx = np.random.randint(0, X_train.shape[0], batch_size)
        imgs = X_train[idx]

        # Sample noise and generate a batch of new images
        noise = np.random.normal(0, 1, (batch_size, self.latent_dim))
        gen_imgs = self.generator.predict(noise)

        # Train the discriminator (real classified as ones and generated as zeros)
        d_loss_real = self.discriminator.train_on_batch(imgs, valid)
```

```
            d_loss_fake =
self.discriminator.train_on_batch(gen_imgs, fake)
            d_loss = 0.5 * np.add(d_loss_real, d_loss_fake)

            # ---------------------
            #  Train Generator
            # ---------------------

            # Train the generator (wants discriminator to
mistake images as real)
            g_loss = self.combined.train_on_batch(noise, valid)
            # Plot the progress
            print ("%d [D loss: %f, acc.: %.2f%%] [G loss: %f]"
% (epoch, d_loss[0], 100*d_loss[1], g_loss))

            # If at save interval => save generated image
samples
            if epoch % save_interval == 0:
                self.save_imgs(epoch)
```

将上述内容组合在一起

在main方法中,将DCGAN类实例化并调用train函数:

```
if __name__ == '__main__':
    dcgan = DCGAN()
    dcgan.train(epochs=4000, batch_size=32, save_interval=50)
```

程序的输出

此处显示了几次迭代后程序的输出。

迭代0次的输出:

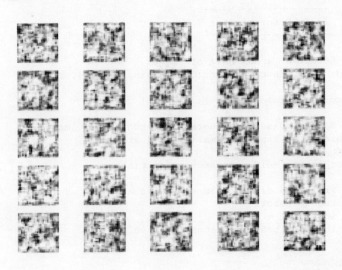

迭代 100 次的输出：

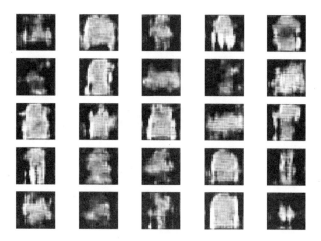

迭代 1 500 次的输出：

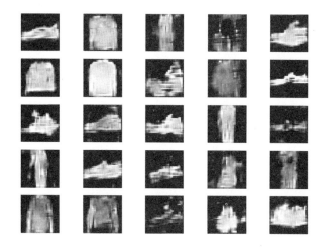

从前面的图片中可以看出，与简单 GAN 和边界搜索 GAN 相比，DCGAN 能用更少的迭代次数获得更加清晰的图像。

模型的平均指标

以下为模型的 G loss、D loss 和 D accuracy：

```
mean d loss:0.6240914929024999
mean g loss:1.2823512871799998
mean d accuracy:65.574609375
```

CHAPTER 7

第 7 章

递归神经网络

本章将介绍以下内容:
- 用于时间序列数据的简单递归神经网络
- 用于时间序列数据的长短期记忆网络
- 长短期记忆内存示例时间序列预测
- 基于长短期记忆的等长输出序列到序列学习

7.1 引言

本章将介绍使用 Keras 创建**递归神经网络（RNN）**的各种方法。首先，需要了解 RNN，从简单的 RNN 开始，然后是**长短期记忆（LSTM）**RNN（由于神经元中的特殊门，这些网络会将状态保留很长一段时间）。

RNN 的重要性

传统神经网络的一个重大缺点是无法保存过去数据间的相互作用，而 RNN 中的循环，能够长时间保存数据信息。如下图所示，神经网络中的 A 块接收输入 x_t，经过运算获得输出 h_t，利用循环使得信息在网络中依次传递。

该图简单显示了递归神经网络的结构:

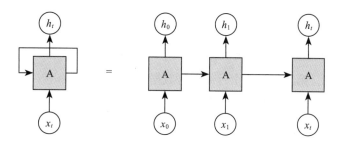

7.2 用于时间序列数据的简单 RNN

本节将介绍如何使用 Keras 的简单 RNN 实现基于历史数据集预测销售额。

> RNN 是一种人工神经网络,其中网络节点之间的连接沿着序列形成有向图,因此可以显示输入随时间序列变化的动态行为。与前馈神经网络不同,RNN 使用其内部状态(也称为**存储器**)来处理输入序列,这使它们适用于诸如未分割的、连接的手写字识别或语音识别之类的任务。

简单的 RNN 实现作为 `keras.layers.Simple RNN` 类中的一部分,如下所示:

```
keras.layers.SimpleRNN(units, activation='tanh',
    use_bias=True,
    kernel_initializer='glorot_uniform',
    recurrent_initializer='orthogonal',
    bias_initializer='zeros',
    kernel_regularizer=None,
    recurrent_regularizer=None,
    bias_regularizer=None,
    activity_regularizer=None,
    kernel_constraint=None,
    recurrent_constraint=None,
    bias_constraint=None,
    dropout=0.0,
    recurrent_dropout=0.0,
    return_sequences=False,
    return_state=False,
    go_backwards=False,
    stateful=False,
    unroll=False)
```

简单 RNN 是全连接 RNN,将其输出反馈并作为下一时刻输入的一部分。接下来使用一个简单的 RNN 进行时间序列预测。

7.2.1 准备工作

保存数据集的文件为：`sale-of-shampoo-over-a-three-ye.csv`。数据有两列，第一列是月份，第二列是每个月的销售数据，如下所示：

```
"Month","Sales of shampoo over a three year period"
"1-01",266.0
"1-02",145.9
"1-03",183.1
"1-04",119.3
"1-05",180.3
"1-06",168.5
"1-07",231.8
```

导入相关的类，如下所示：

```
from pandas import read_csv
from matplotlib import pyplot
from pandas import datetime
```

加载数据集

1. 定义一个解析器将 YY 转换为 YYYY，如下所示：

```
def parser(x):
    return datetime.strptime('200' + x, '%Y-%m')
```

2. 调用 pandas 的 `read_csv` 函数将 `.csv` 文件加载到 pandas 的 `DataFrame`。

> 此时使用的数据解析器是之前已经定义的函数。

3. 调用 `read_csv` 函数：

```
series = read_csv('sales-of-shampoo-over-a-three-ye.csv', header=0,
parse_dates=[0], index_col=0,
                squeeze=True, date_parser=parser)
```

4. 序列加载完成后，打印前几行：

```
print(series.head())
```

输出如下：

```
Month
2001-01-01 266.0
2001-02-01 145.9
```

```
2001-03-01   183.1
2001-04-01   119.3
2001-05-01   180.3
```

5. 使用 `pyplot` 库绘制数据线图：

```
series.plot()
pyplot.show()
```

如下截图为相应的数据线图：

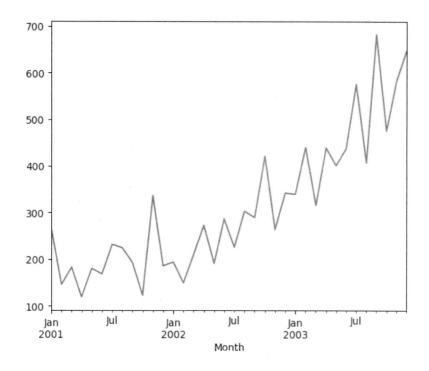

可以看出，销售情况相当不稳定，但总体呈上升趋势。

7.2.2 怎么做

1. 定义简单网络中使用的参数，以及用来储存结果的 `DataFrame`，具体如下所示：

```
n_lag = 1
n_repeats = 30
n_epochs = 1000
n_batch = 4
n_neurons = 3
results = DataFrame()
```

2. 调用 experiment 方法：

```
results['results'] = experiment(series, n_lag, n_repeats, n_epochs,
n_batch, n_neurons)
```

在 experiment() 方法中，通过网络处理数据，如下所示：

```
def experiment(series, n_lag, n_repeats, n_epochs, n_batch,
n_neurons):
    # method details ....
```

3. 得到 series 数据帧的值，如下所示：

```
raw_values = series.values
diff_values = difference(raw_values, 1)
```

raw_values 和 diff values 的输出如下：

```
raw_values :
[266.  145.9 183.1 119.3 180.3 168.5 231.8 224.5 192.8 122.9 336.5
 185.9
 194.3 149.5 210.1 273.3 191.4 287.  226.  303.6 289.9 421.6 264.5
 342.3
 339.7 440.4 315.9 439.3 401.3 437.4 575.5 407.6 682.  475.3 581.3
 646.9]

diff values:

 0  -120.1
 1   37.2
 2  -63.8
 3   61.0
 4  -11.8
 5   63.3
 6   -7.3
 7  -31.7
 8  -69.9
 9  213.6
 10 -150.6
 ...
```

利用每个值减下一个值来计算 diff values，例如 145.9-266.= -120.1。

4. 将时间序列转换为监督值：

```
supervised = timeseries_to_supervised(diff_values, n_lag)
```

timeseries_to_supervised 方法如下所示：

```
def timeseries_to_supervised(data, lag=1):
    df = DataFrame(data)
    columns = [df.shift(i) for i in range(1, lag + 1)]
    columns.append(df)
    df = concat(columns, axis=1)
    return df
```

监督数据帧的输出如下：

```
    0     0
0   NaN  -120.1
1  -120.1  37.2
2   37.2  -63.8
3  -63.8   61.0
4   61.0  -11.8
5  -11.8   63.3
6   63.3   -7.3
7   -7.3  -31.7
8  -31.7  -69.9
9  -69.9  213.6
10  213.6 -150.6
11 -150.6    8.4
12    8.4  -44.8
13  -44.8   60.6
```

5. 我们从监督数据帧中提取 supervised_values，如下代码所示：

```
supervised_values = supervised.values[n_lag:, :]
```

上述代码的输出如下：

```
[[-120.1   37.2]
 [  37.2  -63.8]
 [ -63.8   61. ]
 [  61.   -11.8]
 [ -11.8   63.3]
 [  63.3   -7.3]
 [  -7.3  -31.7]
 [ -31.7  -69.9]
 [ -69.9  213.6]
 [ 213.6 -150.6]
 [-150.6    8.4]
 [   8.4  -44.8]
 [ -44.8   60.6]
 [  60.6   63.2]
 [  63.2  -81.9]
```

6. 将监督值拆分为训练数据帧和测试数据帧，如下所示：

```
train, test = supervised_values[0:-12], supervised_values[-12:]
```

训练数据帧和测试数据帧的输出如下：

```
train :
[[-120.1  37.2]
 [  37.2 -63.8]
 [ -63.8  61. ]
 [  61.  -11.8]
 [ -11.8  63.3]
 [  63.3  -7.3]
 [  -7.3 -31.7]
 [ -31.7 -69.9]
 [ -69.9 213.6]
 [ 213.6 -150.6]
 [-150.6   8.4]
 [   8.4 -44.8]
 [ -44.8  60.6]
 [  60.6  63.2]
 [  63.2 -81.9]

test :
[[  77.8  -2.6]
 [  -2.6 100.7]
 [ 100.7 -124.5]
 [-124.5 123.4]
 [ 123.4 -38. ]
 [ -38.   36.1]
 [  36.1 138.1]
 [ 138.1 -167.9]
 [-167.9 274.4]
 [ 274.4 -206.7]
 [-206.7 106. ]
 [ 106.   65.6]]
```

7. 对训练数据帧和测试数据帧进行缩放处理，如下所示：

```
scaler, train_scaled, test_scaled = scale(train, test)
```

train_scaled数据帧的输出如下：

```
train_scaled
[[-0.80037766  0.04828702]
 [ 0.04828702 -0.496628  ]
 [-0.496628    0.17669274]
 [ 0.17669274 -0.21607769]
 [-0.21607769  0.1891017 ]
 [ 0.1891017  -0.1917993 ]
 [-0.1917993  -0.32344214]
 [-0.32344214 -0.52953871]
 [-0.52953871  1.        ]
 [ 1.
```

```
test_scaled
[[-0.80037766  0.04828702]
 [ 0.04828702 -0.496628  ]
 [-0.496628    0.17669274]
 [ 0.17669274 -0.21607769]
 [-0.21607769  0.1891017 ]
 [ 0.1891017  -0.1917993 ]
 [-0.1917993  -0.32344214]
 [-0.32344214 -0.52953871]
 [-0.52953871  1.        ]
 [ 1.        ]
```

然后，通过神经网络运行缩放后的训练数据帧并计算权重，如下所示：

1. 根据重复次数 `n_repeats` 训练模型。

2. 得到用 `train_scaled [2:,:]` 填充的训练数据帧 `train_trimmed`，具有以下输出：

```
train_trimmed:
 [[-0.496628    0.17669274]
 [ 0.17669274 -0.21607769]
 [-0.21607769  0.1891017 ]
 [ 0.1891017  -0.1917993 ]
 [-0.1917993  -0.32344214]
 [-0.32344214 -0.52953871]
 [-0.52953871  1.        ]
 [ 1.         -0.96493121]
 [-0.96493121 -0.10709469]
 [-0.10709469
```

3. 调用 `fit_rnn(train_trimmed,n_batch,n_epochs,n_neurons)`，它返回 rnn 模型。以下为 `fit_rnn` 实现：

```
def fit_rnn(train, n_batch, nb_epoch, n_neurons):
    X, y = train[:, 0:-1], train[:, -1]
    X = X.reshape(X.shape[0], 1, X.shape[1])
    model = Sequential()
    model.add(SimpleRNN(n_neurons, batch_input_shape=(n_batch, X.shape[1], X.shape[2]),
        stateful=True))
    model.add(Dense(1))
    model.compile(loss='mean_squared_error', optimizer='adam')
    for i in range(nb_epoch):
        model.fit(X, y, epochs=1, batch_size=n_batch, verbose=0, shuffle=False)
        model.reset_states()
    return model
```

首先，通过 `train_trimmed` 得到 X、y。X 和 y 的值为：

```
X :
 [[-0.496628  ]
 [ 0.17669274]
 [-0.21607769]
 [ 0.1891017 ]
 [-0.1917993 ]
 [-0.32344214]
 [-0.52953871]
 [ 1.        ]
 [-0.96493121]
 [-0.10709469]
 [-0.39411923]
 [ 0.17453466]
 [ 0.18856218]
 [-0.59428109]
 [ 0.3633666 ]
 [-0.48152145]
 [ 0.26625303]]
y :
 [ 0.17669274 -0.21607769  0.1891017  -0.1917993  -0.32344214
-0.52953871
  1.         -0.96493121 -0.10709469 -0.39411923  0.17453466  0.18856218
 -0.59428109  0.3633666  -0.48152145  0.26625303 -0.22632857
0.55813326
 -1.          0.26733207]
```

实例化序贯模型

接下来,实例化一个序贯模型并添加以下层:

- 一个简单的 RNN
- 具有一个输出的密集层

以下是创建简单 RNN 的步骤:

1. 模型创建代码如下:

```
model = Sequential()
model.add(SimpleRNN(n_neurons, batch_input_shape=(n_batch,
X.shape[1], X.shape[2]),
    stateful=True))
model.add(Dense(1))
```

2. 通过 `model.compile(..)` 编译模型,设置损失函数和优化器,代码如下所示:

```
model.compile(loss='mean_squared_error', optimizer='adam')
```

使用均方误差(MSE)作为损失函数,Adam 作为优化器。MSE 使用每一组预测值和实际值之间的差的平方和除以 n,其中 n 是总样本大小。

$$MSE = \frac{1}{n}\sum_{i=0}^{i=n}\left(Y_i - \hat{Y}_i\right)^2$$

> **自适应矩估计（Adam）**是另一种计算每个参数的可调学习率的优化方法，它保留每一时刻平方梯度的指数衰减平均值 v_t，梯度的指数衰减平均值 m_t，m_t 和 v_t 计算公式分别如下：

$$m_t = \beta_1 m_{t-1} + (1-\beta_1)g_t$$
$$v_t = \beta_2 v_{t-1} + (1-\beta_2)g_t^2$$

由于 m_t 和 v_t 被初始化为 0 的向量，当衰减率 β_1 和 β_2 为 1 时，它们偏向于 0，因此，利用下式计算偏差调整值 $\hat{m}_t$ 和 $\hat{v}_t$：

$$\widehat{m_t} = \frac{m_t}{1-\beta_1^t}$$

$$\hat{v}_t = \frac{v_t}{1-\beta_2^t}$$

θ 的更新规则如下所示：

$$\theta_{t+1} = \theta_t - \frac{\eta}{\sqrt{v_t + \in}}\hat{m}_t$$

使用以下 Keras 函数实现 Adam：

```
keras.optimizers.Adam(lr=0.001, beta_1=0.9, beta_2=0.999,
epsilon=None, decay=0.0, amsgrad=False)
```

3. 通过在模型运行训练数据来更新模型参数，如下所示：

```
for i in range(nb_epoch):
    model.fit(X, y, epochs=1, batch_size=n_batch, verbose=0,
shuffle=False)
    model.reset_states()
```

4. 返回模型如下：

```
return model
```

接下来，依次完成如下步骤：

1. 改变 `test_reshaped` 测试数据的形状并在模型中运行以获取预测值。

2. 调用 `rnn_model.predict(....)`。

3. 通过 `invert_scale` 和 `invert_difference` 得到 `yhat`。

4. 将预测值存储在预测列表中。

5. 计算 RMSE。

6. 输出每次迭代产生的 RMSE：

```
test_reshaped = test_scaled[:, 0:-1]
test_reshaped = test_reshaped.reshape(len(test_reshaped), 1, 1)
output = rnn_model.predict(test_reshaped, batch_size=n_batch)
predictions = list()
for i in range(len(output)):
    yhat = output[i, 0]
    X = test_scaled[i, 0:-1]
    # invert scaling
    yhat = invert_scale(scaler, X, yhat)
    # invert differencing
    yhat = inverse_difference(raw_values, yhat, len(test_scaled) + 1 - i)
    # store forecast
    predictions.append(yhat)
# report performance
rmse = sqrt(mean_squared_error(raw_values[-12:], predictions))
print('%d) Test RMSE: %.3f' % (r + 1, rmse))
error_scores.append(rmse)
```

7. 模型创建和 `error_scores` 计算的完整代码如下：

```
error_scores = list()
for r in range(n_repeats):
    # fit the model
    train_trimmed = train_scaled[2:, :]
    rnn_model = fit_rnn(train_trimmed, n_batch, n_epochs, n_neurons)
    # forecast test dataset
    test_reshaped = test_scaled[:, 0:-1]
    test_reshaped = test_reshaped.reshape(len(test_reshaped), 1, 1)
    output = lstm_model.predict(test_reshaped, batch_size=n_batch)
    predictions = list()
    for i in range(len(output)):
        yhat = output[i, 0]
        X = test_scaled[i, 0:-1]
        # invert scaling
        yhat = invert_scale(scaler, X, yhat)
        # invert differencing
        yhat = inverse_difference(raw_values, yhat, len(test_scaled) + 1 - i)
        # store forecast
        predictions.append(yhat)
    # report performance
    rmse = sqrt(mean_squared_error(raw_values[-12:], predictions))
    print('%d) Test RMSE: %.3f' % (r + 1, rmse))
    error_scores.append(rmse)
```

8. 获得 30 次迭代过程的 RMSE 分数：

```
1) Test RMSE: 95.838
2) Test RMSE: 75.151
3) Test RMSE: 98.616
4) Test RMSE: 108.205
5) Test RMSE: 83.807
6) Test RMSE: 73.411
...
25) Test RMSE: 86.076
26) Test RMSE: 86.104
27) Test RMSE: 85.667
28) Test RMSE: 74.321
29) Test RMSE: 88.347
30) Test RMSE: 97.868
```

9. RMSE 在迭代过程中的统计结果和绘图显示如下：

```
 results
count 30.000000
mean 86.546032
std 9.338947
min 71.019965
25% 81.406349
50% 85.000358
75% 92.118326
max 108.008031
```

迭代过程中的 RMSE 绘图如下：

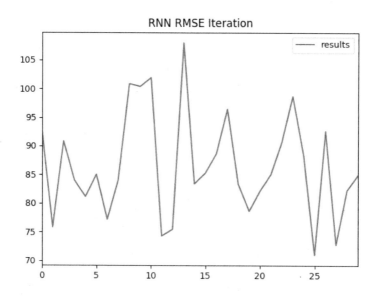

从图中可以看出，在迭代中 RMSE 值在 73 ~ 108 之间变化，并随着时间变化逐渐趋于稳定。

7.3 时间序列数据的 LSTM 网络

本节将介绍什么是 LSTM 网络，以及如何利用它们的长期记忆特性来更好地预测时间序列数据。

7.3.1 LSTM 网络

LSTM 旨在避免长期依赖性问题，它具有长期记忆信息的能力。

所有递归神经网络都具有重复的神经网络模块的链式结构。在标准 RNN 中，该重复模块仅具有一个非常简单的结构，例如单个 tanh 层。LSTM 也具有这种类似的链结构，但重复模块具有不同的结构。

LSTM 有四个层，以非常特殊的方式进行交互，如下图所示：

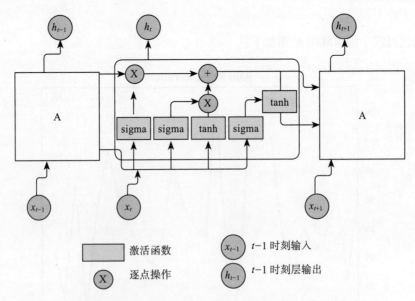

引自：http://colah.github.io/posts/2015-08-Understanding-LSTMs/

比起 LSTM 的工作原理，我们更关注的是如何在 Keras 中使用它。

7.3.2 LSTM 记忆示例

本节中,我们将通过一个简单的例子,了解 LSTM 网络如何记住很久之前的一个过程值。首先输入两个序列,LSTM 将根据第一个序列输入记住要输出的字符,如下所示:

```
seq1 = ['A', 'B', 'C', 'D', 'A']
seq2 = ['Z', 'B', 'C', 'D', 'Z']
```

7.3.3 准备工作

从 `pandas`、`numpy` 和 `keras` 导入相关的 Python 包和类:

```
from pandas import DataFrame
import numpy as np
np.random.seed(1337)
from keras.models import Sequential
from keras.layers import Dense
from keras.layers import LSTM
```

7.3.4 怎么做

深入了解 LSTM 模型是如何能够在预测下一个字符的同时保留上一个序列之前的序列。

编码器

1. 首先,定义一个编码器,它将 `char` 转换为长度为 `91` 的数组中的独热编码值:

```
def encode(pattern, n_unique):
    encoded = list()
    for value in pattern:
        row = [0.0 for x in range(n_unique)]
        index = ord(value)
        row[ord(value)] = 1.0
        encoded.append(row)
    return encoded
```

2. 如下所示将序列分为 `X` 和 `y` 值:

```
def to_xy_pairs(encoded):
    X,y = list(),list()
    for i in range(1, len(encoded)):
        X.append(encoded[i-1])
        y.append(encoded[i])
    return X, y
```

3. 将 X、y 值转换为 LSTM 可以理解的三维矩阵：

```
def to_lstm_dataset(sequence, n_unique):
    # one hot encode
    encoded = encode(sequence, n_unique)
    # convert to in/out patterns
    X,y = to_xy_pairs(encoded)
    # convert to LSTM friendly format
    dfX, dfy = DataFrame(X), DataFrame(y)
    lstmX = dfX.values
    lstmX = lstmX.reshape(lstmX.shape[0], 1, lstmX.shape[1])
    lstmY = dfy.values
    return lstmX, lstmY
```

4. 注意如何从 X、y 和 `lstmX(dfX.values)` 创建 `pandas DataFrame` 并形变为 `model.fit` 的输入：

```
seq1 = ['A', 'B', 'C', 'D', 'A']
seq2 = ['Z', 'B', 'C', 'D', 'Z']
print(ord('z'))
# convert sequences into required data format
#n_unique = len(set(seq1 + seq2))
n_unique = ord('Z') +1
seq1X, seq1Y = to_lstm_dataset(seq1, n_unique)
seq2X, seq2Y = to_lstm_dataset(seq2, n_unique)
```

LSTM 配置和模型

1. 定义 LSTM 配置，根据经验设定期望值以获得所需的输出：

```
# define LSTM configuration
n_neurons = 200
n_batch = 1
n_epoch = 1000
n_features = n_unique
```

2. 以 LSTM 作为其中一个层来定义实际模型：

```
model = Sequential()
model.add(LSTM(n_neurons, batch_input_shape=(n_batch, 1,
n_features), stateful=True))
model.add(Dense(n_unique, activation='sigmoid'))
model.compile(loss='binary_crossentropy', optimizer='adam')
```

这里使用的是一个双层网络，LSTM 作为其中的一个层，随后是一个密集层。

使用的激活函数是 `sigmoid`，损失函数是 `binary_crossentropy`。

3. 使用交叉熵来替代平方误差。当网络的输出为独立假设时（例如，每个节点代表不同的概念），可用交叉熵作为误差度量。节点激活可被理解为每个假设可能为真的概率（或置信度）。输出向量表示概率分布，而交叉熵损失函数是网络认为该分布与它实际之间的距离。

4. 交叉熵在目标值为 0 和 1 的问题中更有用（虽然可以假设输出为 0～1 之间的值），即使在节点饱和（导数渐近于 0）的情况下，交叉熵依旧允许误差改变权重。

5. 使用 Adam 作为梯度下降的优化技术，使用以下 Keras 函数实现 Adam：

```
keras.optimizers.Adam(lr=0.001, beta_1=0.9, beta_2=0.999,
epsilon=None, decay=0.0, amsgrad=False)
```

模型训练

1. 网络编译完成后，进行如下训练：

```
# train LSTM
for i in range(n_epoch):
    model.fit(seq1X, seq1Y, epochs=1, batch_size=n_batch, verbose=1,
shuffle=False)
    model.reset_states()
    model.fit(seq2X, seq2Y, epochs=1, batch_size=n_batch, verbose=0,
shuffle=False)
    model.reset_states()
```

2. 最后一步是对序列一和序列二进行测试，如下所示：

```
# test LSTM on sequence 1
print('Sequence 1')
result = model.predict_classes(seq1X, batch_size=n_batch,
verbose=0)
model.reset_states()
for i in range(len(result)):
    print('X=%s y=%s, yhat=%s' % (seq1[i], seq1[i+1],
chr(result[i])))

# test LSTM on sequence 2
print('Sequence 2')
result = model.predict_classes(seq2X, batch_size=n_batch,
verbose=0)
model.reset_states()
for i in range(len(result)):
    print('X=%s y=%s, yhat=%s' % (seq2[i], seq2[i+1],
chr(result[i])))
```

3. 测试的输出显示 LSTM 模型具有长期记忆的能力，如下所示：

```
Sequence 1
  X=A y=B, yhat=B
  X=B y=C, yhat=C
  X=C y=D, yhat=D
  X=D y=A, yhat=A
Sequence 2
  X=Z y=B, yhat=B
  X=B y=C, yhat=C
  X=C y=D, yhat=D
  X=D y=Z, yhat=Z
```

如下完整代码清单显示了具体逻辑过程。

完整的代码清单

完整代码如下：

```python
from pandas import DataFrame
import numpy as np
np.random.seed(1337)
from keras.models import Sequential
from keras.layers import Dense
from keras.layers import LSTM

# binary encode an input pattern, by converting characters into int
# return a list of binary vectors
def encode(pattern, n_unique):
    encoded = list()
    for value in pattern:
        row = [0.0 for x in range(n_unique)]
        index = ord(value)
        row[ord(value)] = 1.0
        encoded.append(row)
    return encoded


# create input/output pairs of encoded vectors, returns X, y
def to_xy_pairs(encoded):
    X,y = list(),list()
    for i in range(1, len(encoded)):
        X.append(encoded[i-1])
        y.append(encoded[i])
    return X, y
# convert sequence to x/y pairs ready for use with an LSTM
def to_lstm_dataset(sequence, n_unique):
    # one hot encode
    encoded = encode(sequence, n_unique)
    # convert to in/out patterns
    X,y = to_xy_pairs(encoded)
    # convert to LSTM friendly format
    dfX, dfy = DataFrame(X), DataFrame(y)
    lstmX = dfX.values
    lstmX = lstmX.reshape(lstmX.shape[0], 1, lstmX.shape[1])
```

```
    lstmY = dfy.values
    return lstmX, lstmY


seq1 = ['A', 'B', 'C', 'D', 'A']
seq2 = ['Z', 'B', 'C', 'D', 'Z']
print(ord('z'))
# convert sequences into required data format
#n_unique = len(set(seq1 + seq2))
n_unique = ord('Z') +1
seq1X, seq1Y = to_lstm_dataset(seq1, n_unique)
seq2X, seq2Y = to_lstm_dataset(seq2, n_unique)
# define LSTM configuration
n_neurons = 200
n_batch = 1
n_epoch = 1000
n_features = n_unique
# create LSTM
model = Sequential()
model.add(LSTM(n_neurons, batch_input_shape=(n_batch, 1, n_features),
stateful=True))
model.add(Dense(n_unique, activation='sigmoid'))
model.compile(loss='binary_crossentropy', optimizer='adam')
# train LSTM
for i in range(n_epoch):
    model.fit(seq1X, seq1Y, epochs=1, batch_size=n_batch, verbose=1,
shuffle=False)
    model.reset_states()
    model.fit(seq2X, seq2Y, epochs=1, batch_size=n_batch, verbose=0,
shuffle=False)
    model.reset_states()

# test LSTM on sequence 1
print('Sequence 1')
result = model.predict_classes(seq1X, batch_size=n_batch, verbose=0)
model.reset_states()
for i in range(len(result)):
    print('X=%s y=%s, yhat=%s' % (seq1[i], seq1[i+1], chr(result[i])))
# test LSTM on sequence 2
print('Sequence 2')
result = model.predict_classes(seq2X, batch_size=n_batch, verbose=0)
model.reset_states()
for i in range(len(result)):
    print('X=%s y=%s, yhat=%s' % (seq2[i], seq2[i+1], chr(result[i])))
```

7.4 使用 LSTM 进行时间序列预测

本节将学习如何使用 Keras 的 LSTM 实现来基于历史数据集预测销售额。使用之前用于预测洗发水销售额的相同数据集。

7.4.1 准备工作

该数据集在 `sales-of-shampoo-over-a-three-ye.csv` 文件中:

```
"Month","Sales of shampoo over a three year period"
"1-01",266.0
"1-02",145.9
"1-03",183.1
"1-04",119.3
"1-05",180.3
"1-06",168.5
"1-07",231.8
```

导入相关的类:

```python
from pandas import read_csv
from matplotlib import pyplot
from pandas import datetime
```

加载数据集

1. 定义一个解析器将 YY 转换为 YYYY:

```python
def parser(x):
    return datetime.strptime('200' + x, '%Y-%m')
```

2. 调用 `pandas` 中的 `read_csv` 函数将 `.csv` 文件加载到 `DataFrame` 中:

```python
series = read_csv('sales-of-shampoo-over-a-three-ye.csv', header=0,
parse_dates=[0], index_col=0, squeeze=True,
date_parser=parser)
```

3. 使用以下代码,打印前几行:

```python
print(series.head())
```

输出如下所示:

```
Month
2001-01-01  266.0
2001-02-01  145.9
2001-03-01  183.1
2001-04-01  119.3
2001-05-01  180.3
```

4. 使用以下代码打印线图:

```python
series.plot()
pyplot.show()
```

输出如下所示：

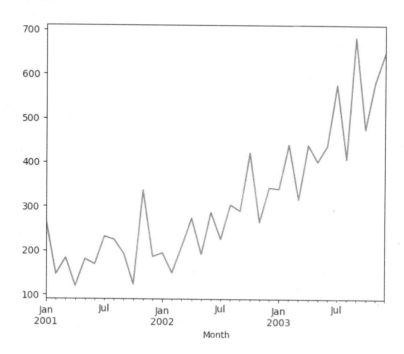

7.4.2 怎么做

1. 定义要在 LSTM 网络中使用的参数，以及用来存储结果的 `DataFrame`：

```
n_lag = 1
n_repeats = 30
n_epochs = 1000
n_batch = 4
n_neurons = 3
results = DataFrame()
```

2. 调用 `experiment` 方法：

```
results['results'] = experiment(series, n_lag, n_repeats, n_epochs,
n_batch, n_neurons)
```

在 `experiment()` 方法中，通过网络处理数据，如下所示：

```
def experiment(series, n_lag, n_repeats, n_epochs, n_batch,
n_neurons):
    # method details ....
```

3. 得到 `series` 数据帧的值，如下所示：

```
raw_values = series.values
diff_values = difference(raw_values, 1)
```

`series` 数据帧的值输出如下：

```
raw_values :
[266.  145.9 183.1 119.3 180.3 168.5 231.8 224.5 192.8 122.9 336.5
 185.9
 194.3 149.5 210.1 273.3 191.4 287.  226.  303.6 289.9 421.6 264.5
 342.3
 339.7 440.4 315.9 439.3 401.3 437.4 575.5 407.6 682.  475.3 581.3
 646.9]
diff values:
 0  -120.1
 1    37.2
 2   -63.8
 3    61.0
 4   -11.8
 5    63.3
 6    -7.3
 7   -31.7
 8   -69.9
 9   213.6
 10 -150.6
 ...
```

差值由每一个数据值减去前一个数据值得到，例如：`145.9-266. = -120.1`。

4. 将时间序列转换为监督值：

```
supervised = timeseries_to_supervised(diff_values, n_lag)
```

5. `timeseries_to_supervised` 方法具体如下：

```
def timeseries_to_supervised(data, lag=1):
    df = DataFrame(data)
    columns = [df.shift(i) for i in range(1, lag + 1)]
    columns.append(df)
    df = concat(columns, axis=1)
    return df
```

监督数据帧的输出如下所示：

```
   0      0
0  NaN   -120.1
1 -120.1   37.2
2  37.2  -63.8
3 -63.8   61.0
4  61.0  -11.8
```

```
 5  -11.8   63.3
 6   63.3   -7.3
 7   -7.3  -31.7
 8  -31.7  -69.9
 9  -69.9  213.6
10  213.6 -150.6
11 -150.6    8.4
12    8.4  -44.8
13  -44.8   60.6
```

6. 从监督数据帧中提取 **supervised_values**,如下所示:

```
supervised_values = supervised.values[n_lag:, :]
```

来自监督数据帧的监督值输出如下:

```
[[-120.1  37.2]
 [  37.2 -63.8]
 [ -63.8  61. ]
 [  61.  -11.8]
 [ -11.8  63.3]
 [  63.3  -7.3]
 [  -7.3 -31.7]
 [ -31.7 -69.9]
 [ -69.9 213.6]
 [ 213.6 -150.6]
 [-150.6   8.4]
 [   8.4 -44.8]
 [ -44.8  60.6]
 [  60.6  63.2]
 [  63.2 -81.9]
```

7. 将监督值拆分为训练数据帧和测试数据帧,如下所示:

```
train, test = supervised_values[0:-12], supervised_values[-12:]
```

训练数据和测试数据集的监督值输出如下:

```
train :
[[-120.1  37.2]
 [  37.2 -63.8]
 [ -63.8  61. ]
 [  61.  -11.8]
 [ -11.8  63.3]
 [  63.3  -7.3]
 [  -7.3 -31.7]
 [ -31.7 -69.9]
 [ -69.9 213.6]
 [ 213.6 -150.6]
 [-150.6   8.4]
```

```
[ 8.4 -44.8]
[ -44.8 60.6]
[ 60.6 63.2]
[ 63.2 -81.9]

test :

[[ 77.8 -2.6]
 [ -2.6 100.7]
 [ 100.7 -124.5]
 [-124.5 123.4]
 [ 123.4 -38. ]
 [ -38. 36.1]
 [ 36.1 138.1]
 [ 138.1 -167.9]
 [-167.9 274.4]
 [ 274.4 -206.7]
 [-206.7 106. ]
 [ 106. 65.6]]
```

8. 对训练数据帧和测试数据帧进行缩放处理，如下所示：

```
scaler, train_scaled, test_scaled = scale(train, test)
```

训练数据帧和测试数据帧的输出如下：

```
train_scaled
[[-0.80037766  0.04828702]
 [ 0.04828702 -0.496628  ]
 [-0.496628    0.17669274]
 [ 0.17669274 -0.21607769]
 [-0.21607769  0.1891017 ]
 [ 0.1891017  -0.1917993 ]
 [-0.1917993  -0.32344214]
 [-0.32344214 -0.52953871]
 [-0.52953871  1.        ]
 [ 1.         
test_scaled
[[-0.80037766  0.04828702]
 [ 0.04828702 -0.496628  ]
 [-0.496628    0.17669274]
 [ 0.17669274 -0.21607769]
 [-0.21607769  0.1891017 ]
 [ 0.1891017  -0.1917993 ]
 [-0.1917993  -0.32344214]
 [-0.32344214 -0.52953871]
 [-0.52953871  1.        ]
 [ 1.         
```

现在，通过神经网络运行缩放后的训练数据集并计算权重，如下所示：

1. 根据重复次数 n_repeats 训练模型

2. 得到用 `train_scaled [2:,:]` 填充的训练数据帧 `train_trimmed`，具有以下输出：

```
train_trimmed:
 [[-0.496628   0.17669274]
 [ 0.17669274 -0.21607769]
 [-0.21607769  0.1891017 ]
 [ 0.1891017  -0.1917993 ]
 [-0.1917993  -0.32344214]
 [-0.32344214 -0.52953871]
 [-0.52953871  1.        ]
 [ 1.         -0.96493121]
 [-0.96493121 -0.10709469]
 [-0.10709469
```

3. 调用 `fit_lstm(train_trimmed,n_batch,n_epochs,n_neurons)`，它返回为 `lstm_model`。

`fit_lstm` 实现如下：

```python
def fit_lstm(train, n_batch, nb_epoch, n_neurons):
    X, y = train[:, 0:-1], train[:, -1]
    X = X.reshape(X.shape[0], 1, X.shape[1])
    model = Sequential()
    model.add(LSTM(n_neurons, batch_input_shape=(n_batch, X.shape[1], X.shape[2]),
        stateful=True))
    model.add(Dense(1))
    model.compile(loss='mean_squared_error', optimizer='adam')
    for i in range(nb_epoch):
        model.fit(X, y, epochs=1, batch_size=n_batch, verbose=0, shuffle=False)
        model.reset_states()
    return model
```

4. 从 `train_trimmed` 得到 X、y。X 和 y 的值如下：

```
X :
 [[-0.496628  ]
 [ 0.17669274]
 [-0.21607769]
 [ 0.1891017 ]
 [-0.1917993 ]
 [-0.32344214]
 [-0.52953871]
 [ 1.        ]
 [-0.96493121]
 [-0.10709469]
 [-0.39411923]
 [ 0.17453466]
```

```
 [ 0.18856218]
 [-0.59428109]
 [ 0.3633666 ]
 [-0.48152145]
 [ 0.26625303]
y:
 [ 0.17669274 -0.21607769 0.1891017 -0.1917993 -0.32344214
-0.52953871
1. -0.96493121 -0.10709469 -0.39411923 0.17453466 0.18856218
 -0.59428109 0.3633666 -0.48152145 0.26625303 -0.22632857
0.55813326
-1. 0.26733207]
```

实例化序贯模型

实例化一个序贯模型并添加以下层:

❑ LSTM

❑ 密集层

以下对具体步骤进行详细描述。

1.具有一个输出的密集层如下:

```
model = Sequential()
model.add(LSTM(n_neurons, batch_input_shape=(n_batch, X.shape[1],
X.shape[2]),
 stateful=True))
model.add(Dense(1))
```

2.使用 `model.compile(..)` 编译带有损失函数和优化器的模型,如下所示:

```
model.compile(loss='mean_squared_error', optimizer='adam')
```

3.以 MSE 作为损失函数,Adam 作为优化器。

4.使用以下 Keras 函数实现 Adam:

```
keras.optimizers.Adam(lr=0.001, beta_1=0.9, beta_2=0.999,
epsilon=None, decay=0.0, amsgrad=False)
```

5.通过运行训练数据来更新模型参数,如下所示:

```
for i in range(nb_epoch):
    model.fit(X, y, epochs=1, batch_size=n_batch, verbose=0,
shuffle=False)
    model.reset_states()
```

6. 返回模型：

```
return model
```

接下来，完成以下步骤：

1. `test_reshaped` 改变测试数据的形状并在模型中运行以获取预测值。

2. 调用 `lstm_model.predict`。

3. 使用 `invert_scale` 和 `inverse_difference` 来找到 `yhat`。

4. 将预测值存储在预测列表中。

5. 计算 RMSE。

6. 输出每次迭代产生的 RMSE。

代码如下。

1. 从改变 `test_reshaped` 的形状开始：

```
test_reshaped = test_scaled[:, 0:-1]
 test_reshaped = test_reshaped.reshape(len(test_reshaped), 1, 1)
 output = lstm_model.predict(test_reshaped, batch_size=n_batch)
 predictions = list()
 for i in range(len(output)):
   yhat = output[i, 0]
   X = test_scaled[i, 0:-1]
   # invert scaling
   yhat = invert_scale(scaler, X, yhat)
   # invert differencing
   yhat = inverse_difference(raw_values, yhat, len(test_scaled) + 1 - i)
   # store forecast
   predictions.append(yhat)
 # report performance
 rmse = sqrt(mean_squared_error(raw_values[-12:], predictions))
 print('%d) Test RMSE: %.3f' % (r + 1, rmse))
 error_scores.append(rmse)
```

2. 模型创建和 `error_scores` 计算的完整代码如下：

```
error_scores = list()
for r in range(n_repeats):
  # fit the model
  train_trimmed = train_scaled[2:, :]
  lstm_model = fit_lstm(train_trimmed, n_batch, n_epochs, n_neurons)
  # forecast test dataset
  test_reshaped = test_scaled[:, 0:-1]
```

```
test_reshaped = test_reshaped.reshape(len(test_reshaped), 1, 1)
output = lstm_model.predict(test_reshaped, batch_size=n_batch)
predictions = list()
for i in range(len(output)):
    yhat = output[i, 0]
    X = test_scaled[i, 0:-1]
    # invert scaling
    yhat = invert_scale(scaler, X, yhat)
    # invert differencing
    yhat = inverse_difference(raw_values, yhat, len(test_scaled) + 1 - i)
    # store forecast
    predictions.append(yhat)
# report performance
rmse = sqrt(mean_squared_error(raw_values[-12:], predictions))
print('%d) Test RMSE: %.3f' % (r + 1, rmse))
error_scores.append(rmse)
```

3. 获得的 RMSE 分数如下：

```
RMSE
 1) Test RMSE: 99.392
 2) Test RMSE: 91.873
 3) Test RMSE: 101.440
 4) Test RMSE: 89.926
 5) Test RMSE: 90.300
 6) Test RMSE: 101.218
 7) Test RMSE: 93.807
 8) Test RMSE: 94.887
 9) Test RMSE: 95.090
10) Test RMSE: 92.210
11) Test RMSE: 98.373
12) Test RMSE: 96.900
13) Test RMSE: 99.465
14) Test RMSE: 91.884
```

4. RMSE 在迭代过程中的统计结果和绘图显示如下：

```
results
count  30.000000
mean   96.420240
std     5.120269
min    88.793766
25%    92.659372
50%    95.393612
75%    99.786859
max   107.698912
```

上述代码的输出截图如下：

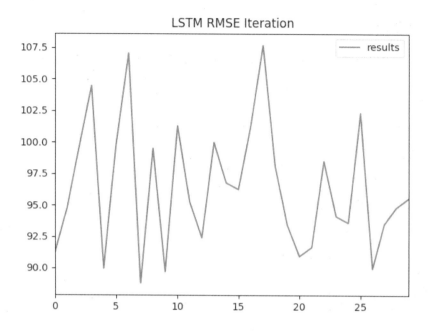

说明

LSTM 的 RMSE 变化小于简单 RNN 网络，但 LSTM（96.42）的平均值高于简单 RNN（86.54）。

7.5　基于 LSTM 的等长输出序列到序列学习

本节将介绍如何使用 LSTM 来预测长度相同或略有不同的值，例如两个数字相减。

7.5.1　准备工作

利用 Keras 和 `six.moves` 依赖项创建 `requirements.txt`。从 `keras`、`numpy` 和 `six.moves` 导入相关类，如下所示：

```
from __future__ import print_function
from keras.models import Sequential
from keras import layers
import numpy as np
import six.moves
```

接下来将介绍如何实现可以处理任意三位数减法的 LSTM 网络。

7.5.2 怎么做

1. 创建一个可以处理编码和解码的字符表，这个类包括如下三个方法：

- __init__()
- encode()
- decode()

2. 代码如下：

```
class CharTable(object):
    def __init__(self, char):
        self.char = sorted(set(char))
        self.char_indices = dict((ch, i) for i, ch in enumerate(self.char))
        self.indices_char = dict((i, ch) for i, ch in enumerate(self.char))

    def encode(self, C, num_rows):
        x = np.zeros((num_rows, len(self.char)))
        for i, ch in enumerate(C):
            x[i, self.char_indices[ch]] = 1
        return x
    def decode(self, x, calc_argmax=True):
        if calc_argmax:
            x = x.argmax(axis=-1)
        return ''.join(self.indices_char[x] for x in x)
```

3. 定义一些实用常量，用于显示最终测试结果、训练数据大小、相减数字的位数以及判断是否训练反例。使查询反向，例如，**12-345** 变为 **543-21**：

```
lass colors:
    ok = '\033[92m'
    fail = '\033[91m'
    close = '\033[0m'

# Parameters for the model and dataset.
TRAINING_SIZE = 50000
DIGITS = 3
REVERSE = True
```

4. 当前输入的最大长度为 3+3+1=7，例如 **300-200 = 7** 位。300 是 3 位数，- 是一位，200 是 3 位数，因此总数是 7 位数。代码如下：

```
MAXLEN = DIGITS + 1 + DIGITS
```

5. 定义字符和字符表实例，如下所示：

```
chars = '0123456789- '
ctable = CharTable(chars)
```

训练数据

1. 通过生成两个随机的三位数字（在这里为 x）生成训练数据，利用减法运算生成 y，如下所示：

```
questions = []
expected = []
seen = set()
print('Generating data...')
while len(questions) < TRAINING_SIZE:
    f = lambda: int(''.join(np.random.choice(list('0123456789'))
                    for i in range(np.random.randint(1, DIGITS +
1))))
    a, b = f(), f()
    # Skip any subtraction questions we've already seen
    # Also skip any such that x-Y == Y-x (hence the sorting).
    key = tuple(sorted((a, b)))
    if key in seen:
        continue
    seen.add(key)
    # Pad the data with spaces such that it is always MAXLEN.
    q = '{}-{}'.format(a, b)
    query = q + ' ' * (MAXLEN - len(q))
    ans = str(a - b)
    # Answers can be of maximum size DIGITS + 1.
    ans += ' ' * (DIGITS + 1 - len(ans))
    if REVERSE:
        # Reverse the query, e.g., '12-345 ' becomes ' 543-21'.
(Note the
        # space used for padding.)
        query = query[::-1]
    questions.append(query)
    expected.append(ans)
print('Total subtraction questions:', len(questions))
```

2. 生成训练数据后，使用字符表对其进行向量化，如下所示：

```
print('Vectorization:')
x = np.zeros((len(questions), MAXLEN, len(chars)), dtype=np.bool)
y = np.zeros((len(questions), DIGITS + 1, len(chars)),
dtype=np.bool)
for i, sentence in enumerate(questions):
 x[i] = ctable.encode(sentence, MAXLEN)
for i, sentence in enumerate(expected):
 y[i] = ctable.encode(sentence, DIGITS + 1)
```

3. 按如下方式对 x 和 y 的值进行重排：

```
indices = np.arange(len(y))
np.random.shuffle(indices)
x = x[indices]
y = y[indices]
```

4.将验证数据设置为10%，这些数据是始终不被训练的。找到需要进行拆分的索引值，例如本例中为45000：

```
split_at = len(x) - len(x) // 10
(x_train, x_val) = x[:split_at], x[split_at:]
(y_train, y_val) = y[:split_at], y[split_at:]
```

5.输出训练和验证数据的形状，如下：

```
Training Data:
(45000, 7, 12)
(45000, 4, 12)
Validation Data:
(5000, 7, 12)
```

模型创建

1.定义模型的超参数：

```
RNN = layers.LSTM
HIDDEN_SIZE = 128
BATCH_SIZE = 128
LAYERS = 1
```

2.定义序贯模型并对其添加各种层，如下所示：

```
print('Build model:')
model = Sequential()
# "Encode" the input sequence using an RNN, producing an output of HIDDEN_SIZE.
# Note: For situation where input sequences have a variable length,
# use input_shape=(None, num_feature).
model.add(RNN(HIDDEN_SIZE, input_shape=(MAXLEN, len(chars))))
model.add(layers.RepeatVector(DIGITS + 1))
for _ in range(LAYERS):
  model.add(RNN(HIDDEN_SIZE, return_sequences=True))
model.add(layers.TimeDistributed(layers.Dense(len(chars))))
model.add(layers.Activation('softmax'))
```

3.`HIDDEN_SIZE`作为第一个LSTM层的输入（本例中为128）具有以下参数的结构：

- 时间步数（`MAXLEN`）
- 特征数量（前一个示例中的`len(chars)`）

它使用重复向量层来获取输入，重复次数为 `DIGITS + 1`，另外添加一个 `LSTM` 层，用于返回隐藏状态更新时间步数（`MAXLEN`），最后一层是 `softmax` 函数 `f`，如下面部分所定义。

> 在数学中，softmax 函数（也称为**归一化指数函数** [1]）是逻辑函数的泛化，用于将任意 K 维向量的实数值 z 压缩为 K 维向量 $\sigma(z)$，压缩后的每个项都分布在 (0，1) 范围内，并且所有项的值加起来为 1。

4. 构建模型后，我们可以看到模型总结，如下：

```
Layer (type) Output Shape Param #
=================================================================
lstm_1 (LSTM) (None, 128) 72192
_____
repeat_vector_1 (RepeatVecto (None, 4, 128) 0
_____
lstm_2 (LSTM) (None, 4, 128) 131584
_____
time_distributed_1 (TimeDist (None, 4, 12) 1548
_____
activation_1 (Activation) (None, 4, 12) 0
=================================================================
Total params: 205,324
Trainable params: 205,324
Non-trainable params: 0
_____
```

模型拟合和预测

1. 迭代地拟合模型，如下所示：

```
for iteration in range(1, 200):
    print()
    print('-' * 50)
    print('Iteration', iteration)
    model.fit(x_train, y_train,
        batch_size=BATCH_SIZE,
        epochs=1,
        validation_data=(x_val, y_val))
    for i in range(10):
        ind = np.random.randint(0, len(x_val))
        rowx, rowy = x_val[np.array([ind])], y_val[np.array([ind])]
```

```
        preds = model.predict_classes(rowx, verbose=0)
        q = ctable.decode(rowx[0])
        correct = ctable.decode(rowy[0])
        guess = ctable.decode(preds[0], calc_argmax=False)
        print('Q', q[::-1] if REVERSE else q, end=' ')
        print('T', correct, end=' ')
        if correct == guess:
            print(colors.ok + '☑' + colors.close, end=' ')
        else:
            print(colors.fail + '☒' + colors.close, end=' ')
        print(guess)
```

上述代码具体执行以下步骤：

1）200次迭代过程中，重复在模型上运行 `x_train` 和 `y_train`，并根据 `x_val` 和 `y_val` 验证模型。

2）取10个随机样本并检查预测的减法值和实际值。

2. 如下所示为上述代码第一次迭代的输出：

```
--------------------------------------------------
Iteration 1
Train on 45000 samples, validate on 5000 samples
Epoch 1/1
2018-06-06 00:36:46.406357: I
tensorflow/core/platform/cpu_feature_guard.cc:140] Your CPU
supports instructions that this TensorFlow binary was not compiled
to use: AVX2 FMA

  128/45000 [..............................] - ETA: 8:04 - loss:
2.4835 - acc: 0.1289
  384/45000 [..............................] - ETA: 2:48 - loss:
2.4789 - acc: 0.1641
.......
44800/45000 [============================>.] - ETA: 0s - loss:
1.8918 - acc: 0.3324
45000/45000 [==============================] - 13s 298us/step -
loss: 1.8907 - acc: 0.3327 - val_loss: 1.6774 - val_acc: 0.3900
Q 226-90 T 136 ☒ 12
Q 30-188 T -158 ☒ -322
Q 57-24 T 33 ☒ -1
Q 878-4 T 874 ☒ 833
Q 78-11 T 67 ☒ 13
Q 452-222 T 230 ☒ -22
Q 859-4 T 855 ☒ 833
Q 969-1 T 968 ☒ 833
Q 722-651 T 71 ☒ -12
Q 983-4 T 979 ☒ 833
```

注意到所有预测都是错误的，但随着迭代次数增加到接近200次，模型的准确率也会逐步提高。

3. 如下所示为迭代 81 次后的输出：

```
--------------------------------------------------
Iteration 80
Train on 45000 samples, validate on 5000 samples
Epoch 1/1
Q 208-4    T 204   ☑ 204
Q 944-826  T 118   ☑ 118
Q 484-799  T -315  ☑ -315
Q 60-408   T -348  ☑ -348
Q 94-742   T -648  ☑ -648
Q 28-453   T -425  ☑ -425
Q 173-63   T 110   ☑ 110
Q 266-81   T 185   ☑ 185
Q 819-57   T 762   ☑ 762
Q 45-904   T -859  ☑ -859
```

该小节给出了很好的关于如何使用 LSTM 进行简单的序列到序列预测的概述，并以任意两个数字的减法为例进行了演示。

CHAPTER 8
第 8 章

使用 Keras 模型进行自然语言处理

本章包括以下内容：
- 词嵌入
- 情感分析

8.1 引言

为什么人类语言如此特别？人类语言或自然语言是一种传达内容的方式，它不由任何形式的物理行为产生，与视觉或其他机器学习任务完全不同。

自然语言处理（NLP）是人工智能（AI）的一种类型，它使得机器能够分析和理解人类语言。目前，NLP 正在开发能够生成和理解自然语言的软件，以便用户能够与自己的计算机进行自然对话。NLP 将 AI 与计算机语言学、计算机科学相结合，以处理人类语言或对话。

NLP 的例子有很多，包括情感分析、聊天机器人、文档分类、单词聚类、机器翻译等，除了很多 NLP 实例外，涉及 NLP 的场景数量更多。本章旨在介绍应用深度学习模型的 NLP 技术，以便你可以轻松地将它们适用于你的数据集，并开发有用的应用程序。

8.2 词嵌入

词嵌入是一种使用密集向量表示单词或文档的 NLP 技术，与使用大型稀疏向量的词袋技术相比，词嵌入通过将数字向量链接到字典中的每个单词，使得单词的语义含义投射到几何空间中，既而通过任何两个向量之间的距离捕获两个相关单词之间的语义关系，由这些向量形成的几何空间被称为**嵌入空间**。

用于学习词嵌入的两种最常见的技术分别是**全局向量单词表示（GloVe）**和**单词到向量表示（Word2vec）**。

接下来，我们将分别利用具有和不具有嵌入层的神经网络处理示例文档。

8.2.1 准备工作

首先，不使用 Keras 中任何预训练的词嵌入。Keras 提供了一个可用于文本或自然语言数据的嵌入层，可以使用 Keras 的 tokenizer API 对输入数据进行数字编码，以便每个词被表示为数值或是整数值。在没有预训练词嵌入的 Keras API 中，嵌入层会使用随机权重进行初始化。

首先创建示例文档和对应的标签，将每个文档分类为正面或负面，代码如下所示：

```
# define documents
documents = ['Well done!',
             'Good work',
             'Great effort',
             'nice work',
             'Excellent!',
             'Weak',
             'Poor effort!',
             'not good',
             'poor work',
             'Could have done better.']
#define class labels
labels = array([1, 1, 1, 1, 1, 0, 0, 0, 0, 0])
```

8.2.2 怎么做

使用 Keras 文本处理 API 对文档进行独热编码。`one_hot` 方法是将分类特征表示为二元向量。首先分类值被映射为整数/数值，然后将整数/数值表示为二元向量，使得除了整数索引处的值之外都是零值。

通常将文档表示为整数值序列，文档中的每个单词都表示为单个整数。

无嵌入

Keras 提供用于文本文档令牌化和编码的 `one_hot()` 函数，它不创建独热编码，而是执行 hashing_trick() 函数，将文本转换为大小固定的哈希空间中的索引序列。

1. 该函数返回文档的整数编码版本：

```
vocab_size = 50
encodeDocuments = [one_hot(doc, vocab_size) for doc in documents]
print(encodeDocuments)
```

上述代码的输出如下：

```
[[1, 39], [37, 40], [21, 19], [5, 40], [16], [36], [8, 19], [25, 37], [8, 40], [25, 44, 39, 26]]
```

其中，Well Done! 和 Good Work 文档分别用向量 [1,39] 和 [37,40] 表示。此外，Could have done better 文档由四个整数的向量 [25,44,39,26] 表示。

2. 然后我们将文档填充到最大长度为 4，如下所示，因为现有向量的最大长度是 4 个整数，如前所示。Keras 库中的 pad_sequences() 函数可用于填充可变长度序列。默认填充值为 0.0，但可以通过 value 参数指定首选值来更改。

3. 可以选择在序列的开头或结尾进行填充，即 pre- 或 post- 序列填充，如下所示：

```
max_length = 4
paddedDocuments = pad_sequences(encodeDocuments, maxlen=max_length,
padding='post')
print(paddedDocuments)
```

上述代码的输出如下：

```
[[ 1 39  0  0]
 [37 40  0  0]
 [21 19  0  0]
 [ 5 40  0  0]
 [16  0  0  0]
 [36  0  0  0]
 [ 8 19  0  0]
 [25 37  0  0]
 [ 8 40  0  0]
 [25 44 39 26]]
```

所有文档都填充 0 达到最大长度为 4。

4. 从 Keras 库中创建一个序贯模型，该模型内部为一系列图层。首先，创建一个新的序贯模型并通过添加层来创建网络拓扑,模型定义后,将模型后端配置为 `TensorFlow`。后端将选择最佳方式来表示网络,从而使得训练过程和预测过程在给定硬件上以最佳方式运行。

5. 如以下代码所示,将嵌入层定义为网络模型的一部分。其中,嵌入词汇表的大小为 50,输入长度为 4。选择八维嵌入空间,这里的模型是二元分类器,而且嵌入层的输出为 4 个向量,每个向量八个维度,一个维度一个单词。我们将其平展为一个 32 元素向量,并传递到密集输出层,最后进行模型拟合和评估。

6. 指定损失函数评估权重集,指定优化器搜索网络中的权重,并指定在训练过程中收集和报告的优化指标。代码如下:

```
model = Sequential()
model.add(Embedding(vocab_size, 8, input_length=max_length))
model.add(Flatten())
model.add(Dense(1, activation='sigmoid'))
model.compile(optimizer='adam', loss='binary_crossentropy',
metrics=['acc'])
print(model.summary())
```

对于给定的分类问题使用相应的对数损失函数,该函数利用 Keras 中的 `binary_crossentropy` 实现。使用梯度下降算法 Adam 进行优化。

> 有关 Adam 优化器(用于随机优化的方法)的更多信息,可参阅 https://arxiv.org/abs/1412.6980v8。

上述代码的输出如下:

```
Layer (type)                 Output Shape              Param #
=================================================================
embedding_1 (Embedding)      (None, 4, 8)              400
_____
flatten_1 (Flatten)          (None, 32)                0
_____
dense_1 (Dense)              (None, 1)                 33
=================================================================
Total params: 433
Trainable params: 433
Non-trainable params: 0
```

现在开始拟合模型。在给定的数据集或文档执行模型,训练过程运行固定的迭代次数,除此之外还可以设置网络权重更新时的评估实例数,即批量大小,利用 `batch_` size 参数进行设置：

```
model.fit(paddedDocuments, labels, epochs=50, verbose=0)
```

7. 最后，在给定文档中对神经网络的性能进行评估，从而获得在训练数据本身的训练准确率。在本章的后面部分，我们将使用训练集和测试集来评估模型的性能，代码如下：

```
loss, accuracy = model.evaluate(paddedDocuments, labels, verbose=0)
print('Accuracy: %f' % (accuracy*100))
```

代码的输出如下：

```
80.000001
```

有嵌入

前面没有涉及如 GloVe 或 Word2vec 的词嵌入功能，现在将使用 Keras 的预训练的词嵌入。还是使用上一节中的文档和标签，代码如下：

```
# define documents
documents = ['Well done!',
             'Good work',
             'Great effort',
             'nice work',
             'Excellent!',
             'Weak',
             'Poor effort!',
             'not good',
             'poor work',
             'Could have done better.']

# define class labels
labels = array([1, 1, 1, 1, 1, 0, 0, 0, 0, 0])
```

Keras 提供了用于准备文本的 tokenizer API，可用于准备多个文本文档。构造一个分词器，然后馈入文本文档或整数编码的文本文档。这里，单词被称为令牌，将文本划分为令牌的方法被描述为令牌化。Keras 中的 text_to_word_sequence API 可将文本拆分为单词列表，如下所示：

```
# use tokenizer and pad
tokenizer = Tokenizer()
tokenizer.fit_on_texts(documents)
vocab_size = len(tokenizer.word_index) + 1
```

```
encodeDocuments = tokenizer.texts_to_sequences(documents)
print(encodeDocuments)
```

代码输出如下:

```
[[6, 2], [3, 1], [7, 4], [8, 1], [9], [10], [5, 4], [11, 3], [5, 1], [12, 13, 2, 14]]
```

`Well Done!` 和 `Good Work` 文档分别用向量 `[6,2]` 和 `[3,1]` 表示, 而 `Could have done better` 文档由四个整数 `[12,13,2,14]` 表示。

与之前的独热编码相比, Keras 的 `tokenizer` 文本 API 更精细, 更适用于生产实际。

如下所示将文档填充到最大长度为 4, Keras 库中的 `pad_sequences()` 函数可用于填充可变长度序列。默认填充值为 `0.0`, 也可以通过 `value` 参数指定首选值来进行填充。

填充可以在序列的开头或结尾使用, 体现为 `pre-` 或 `post-` 序列填充, 如下所示:

```
max_length = 4
paddedDocuments = pad_sequences(encodeDocuments, maxlen=max_length, padding='post')
print(paddedDocuments)
```

代码输出如下:

```
[[ 6  2  0  0] [ 3  1  0  0] [ 7  4  0  0] [ 8  1  0  0] [ 9  0  0  0] [10  0  0  0] [ 5  4  0  0] [11  3  0  0] [ 5  1  0  0] [12 13  2 14]]
```

利用预装的 GloVe 进行词嵌入。GloVe 提供了一套预训练的词嵌入流程, 我们会使用 60 亿词和 100 维的数据训练的 GloVe, 即 `glove.6B.100d.txt`。如果查看文件, 我们可以在每一行上看到一个令牌（单词）, 后面跟着权重（100 个数字）。

在上述步骤中, 我们将整个 GloVe 词嵌入文件作为词嵌入数组的字典加载到内存中。

> 有关 GloVe 的详细信息, 可参阅该网址下的 GloVe 论文: https://nlp.stanford.edu/pubs/glove.pdf。

代码如下:

```
# load glove model
inMemoryGlove = dict()
f = open('/deeplearning-keras/ch08/embeddings/glove.6B.100d.txt')
for line in f:
    values = line.split()
    word = values[0]
```

```python
    coefficients = asarray(values[1:], dtype='float32')
    inMemoryGlove[word] = coefficients
f.close()
print(len(inMemoryGlove))
```

代码输出如下:

```
400000
```

现在,为训练数据集中的每个单词创建一个嵌入矩阵。通过迭代 `Tokenizer.word_index` 中所有单词,并从加载的 GloVe 嵌入中定位嵌入权重向量来完成。

输出是仅针对训练集中单词的权重矩阵,代码如下:

```python
# create coefficient matrix for training data
trainingToEmbeddings = zeros((vocab_size, 100))
for word, i in tokenizer.word_index.items():
    gloveVector = inMemoryGlove.get(word)
    if gloveVector is not None:
        trainingToEmbeddings[i] = gloveVector
```

如前所述,Keras 模型是一系列层。通过创建一个新的序贯模型并通过添加层来创建网络拓扑,模型定义后,将模型后端配置为 `TensorFlow`。后端将选择最佳方式来表示网络,从而使得训练过程和预测过程在给定硬件上以最佳方式运行,代码如下:

```python
model = Sequential()
model.add(Embedding(maxFeatures, 100, weights=[trainingToEmbeddings],
    input_length=max_length, trainable=False))
model.add(Flatten())
model.add(Dense(1, activation='sigmoid'))
model.compile(optimizer='adam', loss='binary_crossentropy',
    metrics=['acc'])
```

对于给定的分类问题使用相应的对数损失函数,该函数利用 Keras 中的 `binary_crossentropy` 实现,使用梯度下降算法 Adam 进行优化。

代码输出结果如下:

```
Layer (type)                 Output Shape              Param #
=================================================================
embedding_1 (Embedding)      (None, 4, 100)            1500
_____
flatten_1 (Flatten)          (None, 400)               0
_____
dense_1 (Dense)              (None, 1)                 401
=================================================================
Total params: 1,901
```

```
Trainable params: 401
Non-trainable params: 1,500
```

现在开始拟合模型。在给定数据集或文档上执行模型，训练过程运行固定的迭代次数，除此之外还可以设置网络权重更新时的评估实例数，即**批量大小**，利用 `batch_size` 参数进行设置，代码如下：

```
model.fit(paddedDocuments, labels, epochs=50, verbose=0)
```

最后，在给定文档中对神经网络的性能进行评估，从而获得在训练数据本身的训练准确率。同样，在本章的后一部分，我们将使用训练数据集和测试数据集来评估模型的性能，如下所示：

```
loss, accuracy = model.evaluate(paddedDocuments, labels, verbose=0)
print('Accuracy: %f' % (accuracy * 100))
```

上述代码的输出如下：

```
100.000000
```

8.3 情感分析

随着技术对企业的进步起着越来越重要的作用，情感分析正成为各种商业案例中的常用工具，企业使用情绪分析技术得到客户对企业业务、产品和相关主题的感受。

情感分析基本上是用于识别并分类一段文本或语料库中所表达的情感的一种方法，为了确定他人对特定主题、产品等的态度是积极的、消极的还是中立的，使用 NLP 对应地将文档分类为正面、中性或负面。

本节将介绍如何开发用于情感分析的深度学习模型，包括：

- 如何在 Keras 中预处理和加载数据集
- 如何使用词嵌入
- 如何开发一个用于情感分析的大型神经网络模型

8.3.1 准备工作

加载数据集并计算相关属性。首先从加载情感数据集开始，提取文本和相应的情感标签，仅保留数据集中必要的列。

> **关于数据集**
>
> 这些数据最初来自 Crowdflower's data for everyone 数据库（https://www.figure-eight.com/data-for-everyone/）。
>
> 最初的数据来源是支持者浏览了数以万计关于俄亥俄州 8 月初共和党（GOP）辩论的推文后，对支持者进行的情感分析和数据分类。支持者被问及推文是否相关，提到了哪个候选人，提到了什么主题和针对一些特定推文的情感。最终得到的数据集已经删除掉了一些不相关消息，并作为代码存储库的一部分直接可用。

接下来，放弃中性情感，因为我们的目标是区分正面推文和负面推文，然后对推文过滤，只保留有效的文本和单词，最后，将最大特征数量值定义为 2 000，并使用分词器进行向量化处理将文本转换为序列，将该序列作为网络的输入：

```
# read input document
X = pd.read_csv('/deeplearning-keras/ch08/sentiment-analysis/Sentiment.csv')
X = X[['text', 'sentiment']]
X = X[X.sentiment != 'Neutral']
X['text'] = X['text'].apply(lambda x: x.lower())
X['text'] = X['text'].apply((lambda x: re.sub('[^a-zA-Z0-9\s]', '', x)))

for idx, row in X.iterrows():
    row[0] = row[0].replace('rt', ' ')

print(X)
```

示例文本输出如下，其中包含文本和情绪列：

```
index    text                                              sentiment
1  rt scottwalker didnt catch the full gopdebate ... Positive
3  rt robgeorge that carly fiorina is trending h... Positive
4  rt danscavino gopdebate w realdonaldtrump deli... Positive
5  rt gregabbott_tx tedcruz on my first day i wil... Positive
6  rt warriorwoman91 i liked her and was happy wh... Negative
8  deer in the headlights rt lizzwinstead ben car... Negative
9  rt nancyosborne180 last nights debate proved i... Negative
10 jgreendc realdonaldtrump in all fairness billc... Negative
11 rt waynedupreeshow just woke up to tweet this ... Positive
12 me reading my familys comments about how great... Negative
```

8.3.2 怎么做

Keras 中的 `tokenizer` API 有几种方法可以帮助我们处理文本以适用于神经网络模型，这里使用 `fit_on_texts` 方法，另外使用 `word_index` 查看单词索引。

1. Keras 为我们提供了 `text_to_word_sequence` API，可用于将文本拆分为单词列表：

```
# use tokenizer and pad
maxFeatures = 2000
tokenizer = Tokenizer(num_words=maxFeatures, split=' ')
tokenizer.fit_on_texts(X['text'].values)
# maxFeatures = len(tokenizer.word_index) + 1
encodeDocuments = tokenizer.texts_to_sequences(X['text'].values)
```

上述代码的输出如下：

```
[[363, 122, 1, 703, 2, 39, 58, 237, 37, 210, 6, 174, 1761, 12,
1324, 1409, 743], [16, 284, 252, 5, 821, 102, 167, 26, 136, 6, 1,
173, 12, 2, 233, 724, 17], so on.
```

2. 然后使用 Keras 库中的 `pad_sequences()` 函数将文档填充到最大长度 `29`，如下所示。其中默认填充值为 `0.0`，但可以通过 `value` 参数指定首选值来进行更改。

3. 填充可以在序列的开头或结尾使用，体现为 `pre-` 或 `post-` 序列填充，如下所示：

```
max_length = 29
paddedDocuments = pad_sequences(encodeDocuments, maxlen=max_length,
    padding='post')
```

4. 接下来，如前所述，利用预装的 GloVe 进行词嵌入。基本上，`GloVe` 提供了一套预训练的词嵌入流程，我们会使用 60 亿词和 100 维的数据训练的 GloVe，即 `glove.6B.100d.txt`。如果查看文件，我们可以在每一行上看到一个令牌（单词），后面跟着权重（100 个数字）。

5. 因此，我们将整个 GloVe 词嵌入文件作为词嵌入数组的字典加载到内存中：

```
# load glove model
inMemoryGlove = dict()
f = open('/deeplearning-keras/ch08/embeddings/glove.6B.100d.txt')
for line in f:
    values = line.split()
    word = values[0]
```

```
        coefficients = asarray(values[1:], dtype='float32')
        inMemoryGlove[word] = coefficients
f.close()
print(len(inMemoryGlove))
```

上述代码的输出如下:

```
400000
```

6. 将标签转换为 1 和 0，分别对应正面值和负面值：

```
# split data
labels = []
for i in X['sentiment']:
    if i == 'Positive':
        labels.append(1)
    else:
        labels.append(0)

labels = array(labels)
```

学习模型旨在对以后未知数据集做出良好的预测。但问题是，完全在现有的数据集上创建模型，如何获得未知数据？一种方法是将数据集分成两组，称为**训练集**和**测试集**，作为原始数据集的子集。

7. 数据集通常被分成训练集和测试集。训练集包含特征向量和相应的输出或标签，模型用于学习以泛化到其他数据集。创建测试数据集（或子集）以测试模型的预测是否精准。从 scikit-learn 的 `model_selection` 子库中导入 `train_test_split` 函数以拆分训练集和测试集，如下所示：

```
X_train, X_test, Y_train, Y_test =
train_test_split(paddedDocuments,labels, test_size = 0.33,
random_state = 42)
print(X_train.shape,Y_train.shape)
print(X_test.shape,Y_test.shape)
```

上述代码的输出如下：

```
(7188, 29) (7188,) (3541, 29) (3541,)
```

8. 现在，为训练数据集中的每个单词创建一个嵌入矩阵。通过迭代 `Tokenizer.word_index` 中的所有单词并从加载的 GloVe 中定位嵌入权重向量来实现这一点。

输出是权重矩阵，但仅适用于训练集中的单词，代码如下：

```python
# create coefficient matrix for training data
trainingToEmbeddings = zeros((maxFeatures, 100))
for word, i in tokenizer.word_index.items():
    if i < 2001:
        gloveVector = inMemoryGlove.get(word)
        if gloveVector is not None:
            trainingToEmbeddings[i] = gloveVector
```

9. 如前所述，Keras 模型是一系列层。通过创建一个新的序贯模型并通过添加层来创建网络拓扑，定义好模型之后，使用 Tensorflow 作为后端进行编译，此后端选择最佳方式来表示网络，从而使得训练过程和预测过程在给定硬件上以最佳方式运行，代码如下：

```python
model = Sequential()
model.add(Embedding(maxFeatures, 100,
    weights=[trainingToEmbeddings], input_length=max_length,
    trainable=False))
model.add(Flatten())
model.add(Dense(1, activation='sigmoid'))
model.compile(optimizer='adam', loss='binary_crossentropy',
    metrics=['acc'])
print(model.summary())
```

10. 对于给定的分类问题使用相应的对数损失函数，该函数利用 Keras 中的 `binary_crossentropy` 实现。使用梯度下降算法 Adam 进行优化。

输出如下：

```
Layer (type)                 Output Shape              Param #
=================================================================
embedding_1 (Embedding)      (None, 29, 100)           200000
_____
flatten_1 (Flatten)          (None, 2900)              0
_____
dense_1 (Dense)              (None, 1)                 2901
=================================================================
Total params: 202,901
Trainable params: 2,901
Non-trainable params: 200,000
```

11. 现在来拟合这个模型。在给定的数据集或文档上执行模型。训练过程运行固定的迭代次数。除此之外还可以设置网络权重更新时的评估实例数，即批量大小，利用 `batch_size` 参数进行设置，如下所示：

```python
batch_size = 32
model.fit(X_train, Y_train, epochs=50, batch_size=batch_size,
    verbose=0)
```

12. 最后，在给定文档中对神经网络的性能进行评估。通过以下代码可知完成这个情感分析任务的准确率约为 81%：

```
loss, accuracy = model.evaluate(X_test, Y_test, verbose=0)
print('Accuracy: %f' % (accuracy * 100))
```

准确率输出如下：

```
81.191754
```

8.3.3 完整代码清单

完整代码清单如下：

```
# imports from Keras and Sklearn
import csv
from numpy import array, asarray, zeros
from keras.preprocessing.text import one_hot, Tokenizer
from keras.preprocessing.sequence import pad_sequences
from keras.models import Sequential
from keras.layers import Dense, SpatialDropout1D, LSTM
from keras.layers import Flatten
from keras.layers.embeddings import Embedding
import pandas as pd
import re
from sklearn.model_selection import train_test_split

# read input document - Sentiment.csv (part of repository)
X = pd.read_csv('/deeplearning-keras/ch08/sentiment-
analysis/Sentiment.csv')
X = X[['text', 'sentiment']]
X = X[X.sentiment != 'Neutral']
X['text'] = X['text'].apply(lambda x: x.lower())
X['text'] = X['text'].apply((lambda x: re.sub('[^a-zA-Z0-9\s]', '', x)))
print(X)

for idx, row in X.iterrows():
    row[0] = row[0].replace('rt', ' ')

# use tokenizer and pad_sequences for processing the input documents
maxFeatures = 2000
tokenizer = Tokenizer(num_words=maxFeatures, split=' ')
tokenizer.fit_on_texts(X['text'].values)
encodeDocuments = tokenizer.texts_to_sequences(X['text'].values)
print(encodeDocuments)

max_length = 29
paddedDocuments = pad_sequences(encodeDocuments, maxlen=max_length,
padding='post')
```

```python
# load glove model 'glove.6B.100d'
inMemoryGlove = dict()
f = open('/deeplearning-keras/ch08/embeddings/glove.6B.100d.txt')
for line in f:
    values = line.split()
    word = values[0]
    coefficients = asarray(values[1:], dtype='float32')
    inMemoryGlove[word] = coefficients
f.close()
print(len(inMemoryGlove))

# convert label's Positive and Negative to 1 and 0 respectively
labels = []
for i in X['sentiment']:
    if i == 'Positive':
        labels.append(1)
    else:
        labels.append(0)

labels = array(labels)

# split data into training and testing sets
X_train, X_test, Y_train, Y_test = train_test_split(paddedDocuments,labels,
test_size = 0.33, random_state = 42)
print(X_train.shape,Y_train.shape)
print(X_test.shape,Y_test.shape)

# create coefficient matrix for training data
trainingToEmbeddings = zeros((maxFeatures, 100))
for word, i in tokenizer.word_index.items():
    if i < 2001:
        gloveVector = inMemoryGlove.get(word)
        if gloveVector is not None:
            trainingToEmbeddings[i] = gloveVector

# create sequential model and add layers
model = Sequential()
model.add(Embedding(maxFeatures, 100, weights=[trainingToEmbeddings],
input_length=max_length, trainable=False))
model.add(Flatten())
model.add(Dense(1, activation='sigmoid'))
model.compile(optimizer='adam', loss='binary_crossentropy',
metrics=['acc'])
print(model.summary())

# finally fit the model on the training dataset
batch_size = 32
model.fit(X_train, Y_train, epochs=50, batch_size=batch_size, verbose=0)

# evaluate loss and accuracy on the testing dataset
loss, accuracy = model.evaluate(X_test, Y_test, verbose=0)
print('Accuracy: %f' % (accuracy * 100))
```

CHAPTER 9
第 9 章

基于 Keras 模型的文本摘要

本章将介绍以下内容：
- 评论的文本摘要

9.1 引言

文本摘要是**自然语言处理（NLP）**中用于生成参考文档精简摘要的方法。当前手动生成大型文档的摘要仍是一项非常困难的任务，所以使用机器学习进行文本摘要仍是一个活跃的研究课题，本章末尾提供了一些相关的参考资料。在进一步研究如何生成文本摘要之前，首先给出摘要的定义：摘要是从一个或多个文本生成的文本输出，它以较短的形式传达原始文本的相关信息，而自动文本摘要的目标是根据语义对原文本进行缩减。近年来出现了很多使用 NLP 技术的自动文本摘要方法，并且已经在各种领域中广泛应用，其中包括用于创建摘要的搜索引擎以及新闻网站中生成新闻主题综述的预览摘要，这种摘要通常作为标题来辅助用户浏览。

为了有效地进行文本摘要，深度学习模型需要能够理解文档并提取重要信息，这些过程非常复杂，而且难度随着文档长度的增加而增加。

9.2 评论的文本摘要

接下来针对全球最大的电子商务平台亚马逊上精美食品的产品评论创建相关摘要，

其中除了产品信息、用户信息、排名和纯文本评论之外，还包括其他类别的评论。我们通过定义编码器—解码器**递归神经网络（RNN）**架构来开发基本的字符级**序列到序列**（seq2seq）模型。

数据集包括以下内容：
- 568 454 条评论
- 256 059 个用户
- 74 258 个产品

> 可以在以下网址中找到本章使用的数据集：https://www.kaggle.com/snap/amazon-fine-food-reviews/。

9.2.1 怎么做

本节将开发一个模型流水线和编码器—解码器架构，为给定的一组评论创建相关的摘要。模型流水线使用 Keras 函数 API 编写的 RNN 模型和各种数据操作的库。

利用编码器—解码器架构构建用于序列预测的 RNN 模型。其中，编码器用于读取完整输入序列并将其编码为固定长度向量的内部表示，具体为上下文向量；解码器则从编码器读取编码的输入序列并生成输出序列。最常见的是基于双向 RNN 的编码器，例如 LSTM。

数据处理

至关重要的是，以正确的数据作为神经网络的输入并对其进行训练和验证。首先需要确保数据具有合理的尺寸和格式，并包含有意义的特征，从而得到更好、更一致的输出结果。

采用以下工作流程进行数据预处理：

1. 使用 `pandas` 加载数据集
2. 将数据集拆分为输入、输出变量以进行机器学习
3. 对输入变量进行预处理变换
4. 总结数据以显示其变化

具体步骤：

1. 首先导入重要的包和数据集。使用 pandas 库加载数据并查看数据集的形状，数据集包含 10 个特征维度和 500 万个数据点：

```
import pandas as pd
import re
from nltk.corpus import stopwords
from pickle import dump, load

reviews = pd.read_csv("/deeplearning-keras/ch09/summarization/Reviews.csv")
print(reviews.shape)
print(reviews.head())
print(reviews.isnull().sum())
```

输出如下所示：

```
(568454, 10) Id 0
ProductId 0
UserId 0
ProfileName 16
HelpfulnessNumerator 0
HelpfulnessDenominator 0
Score 0
Time 0
Summary 27
Text 0
```

2. 删除空值和不需要的特征维度，如下面的代码段所示：

```
reviews = reviews.dropna()
reviews = reviews.drop(['Id','ProductId','UserId','ProfileName','HelpfulnessNumerator','HelpfulnessDenominator', 'Score','Time'], 1)
reviews = reviews.reset_index(drop=True) print(reviews.head())
for i in range(5):
    print("Review #",i+1)
    print(reviews.Summary[i])
    print(reviews.Text[i])
    print()
```

输出如下：

```
Summary Text
 0 Good Quality Dog Food I have bought several of the Vitality canned d...
 1 Not as Advertised Product arrived labeled as Jumbo Salted Peanut...
 2 "Delight," says it all This is a confection that has been around a fe...
 3 Cough Medicine If you are looking for the secret ingredient i...
```

```
Review # 1
Not as Advertised - Product arrived labeled as Jumbo Salted
Peanuts...the peanuts were actually small sized unsalted. Not sure
if this was an error or if the vendor intended to represent the
product as "Jumbo".
Review # 2
"Delight" says it all - This is a confection that has been around a
few centuries. It is a light, pillowy citrus gelatin with nuts - in
this case, Filberts. And it is cut into tiny squares and then
liberally coated with powdered sugar. And it is a tiny mouthful of
heaven. Not too chewy, and very flavorful. I highly recommend this
yummy treat. If you are familiar with the story of C.S. Lewis' "The
Lion, The Witch, and The Wardrobe" - this is the treat that seduces
Edmund into selling out his Brother and Sisters to the Witch.
Review # 3
Cough Medicine - If you are looking for the secret ingredient in
Robitussin I believe I have found it. I got this in addition to the
Root Beer Extract I ordered (which was good) and made some cherry
soda. The flavor is very medicinal.
```

根据定义，缩略是省略两个单词之间的一些字母和符号的简化形式。

> 从 http://stackoverflow.com/questions/19790188/expanding-english-language-contractions-in-python 中获取缩略列表。

3. 用较完整的形式替换缩略词，如下所示：

```
contractions = {
 "ain't": "am not",
 "aren't": "are not",
 "can't": "cannot",
 "can't've": "cannot have",
 "'cause": "because",
 "could've": "could have",
 "couldn't": "could not",
 "couldn't've": "could not have",
 "didn't": "did not",
 "doesn't": "does not",
 "don't": "do not",
 "hadn't": "had not",
 "hadn't've": "had not have",
 "hasn't": "has not",
 "haven't": "have not",
 "he'd": "he would",
 "he'd've": "he would have",
```

4. 通过替换缩略词和删除停止词来简化文本文档：

```
def clean_text(text, remove_stopwords=True):
    # Convert words to lower case
```

```
        text = text.lower()

    if True:
        text = text.split()
        new_text = []
        for word in text:
            if word in contractions:
                new_text.append(contractions[word])
            else:
                new_text.append(word)
        text = " ".join(new_text)

    text = re.sub(r'https?:\/\/.*[\r\n]*', '', text,
flags=re.MULTILINE)
    text = re.sub(r'\<a href', ' ', text)
    text = re.sub(r'&amp;', '', text)
    text = re.sub(r'[_"\-;%()|+&=*%.,!?:#$@\[\]/]', ' ', text)
    text = re.sub(r'<br />', ' ', text)
    text = re.sub(r'\'', ' ', text)

    if remove_stopwords:
        text = text.split()
        stops = set(stopwords.words("english"))
        text = [w for w in text if not w in stops]
        text = " ".join(text)

    return text
```

5. 删除例如停止词这类不需要的字符，并确保更换缩略词。从**自然语言工具包**（**NLTK**）中获取停止词列表，帮助我们在段落中分割句子、分割单词和识别词性。导入工具包的指令如下：

```
import nltk
nltk.download('stopwords')
```

6. 清理摘要，代码如下所示：

```
# Clean the summaries and texts
clean_summaries = []
for summary in reviews.Summary:
    clean_summaries.append(clean_text(summary,
remove_stopwords=False))
print("Summaries are complete.")

clean_texts = []
for text in reviews.Text:
    clean_texts.append(clean_text(text))
print("Texts are complete.")
```

7. 最后，将所有评论保存到 pickle 文件中。使用 pickle 序列化对象，将它们保存到文件中，以便以后再次加载到程序中：

```
stories = list()
for i, text in enumerate(clean_texts):
 stories.append({'story': text, 'highlights': clean_summaries[i]})

# save to file
dump(stories, open('/deeplearning-
keras/ch09/summarization/review_dataset.pkl', 'wb'))
```

编码器—解码器结构

开发一个用于文本摘要的字符级 seq2seq 模型，当然也可以使用文本处理中常见的单词级模型，本书具体介绍字符级模型的使用。编码器和解码器架构是一种用于创建序列预测 RNN 的方法。编码器读取整个输入序列并将其编码为内部表示，通常是固定长度的向量，称为上下文向量。解码器从编码器读取编码的输入序列并产生输出序列。

编码器—解码器架构主要由两个模型组成：一个读取输入序列并将其编码为固定长度向量，另一个解码固定长度向量并输出预测序列。这种架构专为 seq2seq 问题而设计。

1. 首先，定义超参数，例如批量大小、训练次数和训练的样本数：

```
batch_size = 64
epochs = 110
latent_dim = 256
num_samples = 10000
```

2. 接下来，从 `pickle` 文件加载评论数据集：

```
stories = load(open('/deeplearning-
keras/ch09/summarization/review_dataset.pkl', 'rb'))
print('Loaded Stories %d' % len(stories))
print(type(stories))
```

输出如下：

```
Loaded Stories 568411
```

3. 然后对数据进行向量化：

```
input_texts = []
 target_texts = []
 input_characters = set()
 target_characters = set()
 for story in stories:
     input_text = story['story']
     for highlight in story['highlights']:
```

```
            target_text = highlight

    # We use "tab" as the "start sequence" character
    # for the targets, and "\n" as "end sequence" character.
    target_text = '\t' + target_text + '\n'
    input_texts.append(input_text)
    target_texts.append(target_text)
    for char in input_text:
        if char not in input_characters:
            input_characters.add(char)
    for char in target_text:
        if char not in target_characters:
            target_characters.add(char)

input_characters = sorted(list(input_characters))
target_characters = sorted(list(target_characters))
num_encoder_tokens = len(input_characters)
num_decoder_tokens = len(target_characters)
max_encoder_seq_length = max([len(txt) for txt in input_texts])
max_decoder_seq_length = max([len(txt) for txt in target_texts])

print('Number of samples:', len(input_texts))
print('Number of unique input tokens:', num_encoder_tokens)
print('Number of unique output tokens:', num_decoder_tokens)
print('Max sequence length for inputs:', max_encoder_seq_length)
print('Max sequence length for outputs:', max_decoder_seq_length)
```

输出如下：

```
Number of samples: 568411
Number of unique input tokens: 84
Number of unique output tokens: 48
Max sequence length for inputs: 15074
Max sequence length for outputs: 5
```

4. 创建一个通用函数来定义编码器—解码器 RNN：

```
def define_models(n_input, n_output, n_units):
    # define training encoder
    encoder_inputs = Input(shape=(None, n_input))
    encoder = LSTM(n_units, return_state=True)
    encoder_outputs, state_h, state_c = encoder(encoder_inputs)
    encoder_states = [state_h, state_c]
    # define training decoder
    decoder_inputs = Input(shape=(None, n_output))
    decoder_lstm = LSTM(n_units, return_sequences=True,
return_state=True)
    decoder_outputs, _, _ = decoder_lstm(decoder_inputs,
initial_state=encoder_states)
    decoder_dense = Dense(n_output, activation='softmax')
    decoder_outputs = decoder_dense(decoder_outputs)
```

```
    model = Model([encoder_inputs, decoder_inputs],
decoder_outputs)
    # define inference encoder
    encoder_model = Model(encoder_inputs, encoder_states)
    # define inference decoder
    decoder_state_input_h = Input(shape=(n_units,))
    decoder_state_input_c = Input(shape=(n_units,))
    decoder_states_inputs = [decoder_state_input_h,
decoder_state_input_c]
    decoder_outputs, state_h, state_c =
decoder_lstm(decoder_inputs,  initial_state=decoder_states_inputs)
    decoder_states = [state_h, state_c]
    decoder_outputs = decoder_dense(decoder_outputs)
    decoder_model = Model([decoder_inputs] + decoder_states_inputs,
[decoder_outputs] + decoder_states)
    # return all models
    return model, encoder_model, decoder_model
```

训练

1. 执行训练程序，使用 **rmsprop** 优化器和 **categorical_crossentropy** 损失函数：

```
# Run training
 model.compile(optimizer='rmsprop',
loss='categorical_crossentropy')
 model.fit([encoder_input_data, decoder_input_data],
decoder_target_data,
          batch_size=batch_size,
          epochs=epochs,
          validation_split=0.2)
 # Save model
 model.save('/deeplearning-keras/ch09/summarization/model2.h5')
```

代码输出如下：

```
64/800 [=>............................] - ETA: 22:05 - loss: 2.1460
128/800 [===>..........................] - ETA: 18:51 - loss: 2.1234
192/800 [======>.......................] - ETA: 16:36 - loss: 2.0878
256/800 [========>.....................] - ETA: 14:38 - loss: 2.1215
320/800 [==========>...................] - ETA: 12:47 - loss: 1.9832
384/800 [=============>................] - ETA: 11:01 - loss: 1.8665
448/800 [================>.............] - ETA: 9:17 - loss: 1.7547
512/800 [==================>...........] - ETA: 7:35 - loss: 1.6619
576/800 [=====================>........] - ETA: 5:53 - loss: 1.5820
```

```
512/800 [=================>..........] - ETA: 7:19 - loss: 0.7519
576/800 [=====================>........] - ETA: 5:42 - loss: 0.7493
640/800 [======================>......] - ETA: 4:06 - loss: 0.7528
704/800 [=========================>....] - ETA: 2:28 - loss: 0.7553
768/800 [============================>..] - ETA: 50s - loss: 0.7554
```

2. 对于结果的推出，使用以下方法：

```
# generate target given source sequence
 def predict_sequence(infenc, infdec, source, n_steps,
cardinality):
    # encode
    state = infenc.predict(source)
    # start of sequence input
    target_seq = array([0.0 for _ in
range(cardinality)]).reshape(1, 1, cardinality)
    # collect predictions
    output = list()
    for t in range(n_steps):
        # predict next char
        yhat, h, c = infdec.predict([target_seq] + state)
        # store prediction
        output.append(yhat[0,0,:])
        # update state
        state = [h, c]
        # update target sequence
        target_seq = yhat
    return array(output)
```

输出如下：

```
Review(1): The coffee tasted great and was at such a good price! I
highly recommend this to everyone!
 Summary(1): great coffee
Review(2): This is the worst cheese that I have ever bought! I will
never buy it again and I hope you won't either!
 Summary(2): omg gross gross
Review(3): love individual oatmeal cups found years ago sam quit
selling sound big lots quit selling found target expensive buy
individually trilled get entire case time go anywhere need water
microwave spoon to know quaker flavor packets
 Summary(3): love it
```

9.2.2 参考资料

A Deep Reinforced Model for Abstractive Summarization: https://arxiv.org/abs/1705.04304

State-of-the-art abstractive summarization: https://web.stanford.edu/class/cs224n/reports/6878681.pdf

Taming Recurrent Neural Networks for Better Summarization: http://www.abigailsee.com/2017/04/16/taming-rnns-for-better-summarization.html

CHAPTER 10

第 10 章

强化学习

本章将介绍以下内容：

- 使用 Keras 进行《CartPole》游戏
- 使用竞争 DQN 算法进行《CartPole》游戏

10.1 引言

强化学习是机器学习的一个分支，其中 AI 智能体通过与环境交互来从环境中学习并提高性能，而且该智能体的学习方式是反复试错而不是人为监督。下图说明了 AI 智能体如何对环境起作用并在每个操作后收到反馈。反馈由两部分组成：奖励和下一个环境状态。奖励由人类定义：

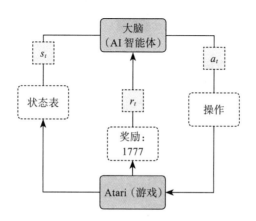

谷歌 DeepMind 在 2013 年发表了一篇名为《 Playing Atari with Deep Reinforcement Learning 》的论文，文中介绍了一种称为 Deep Q Network（DQN）的新算法。该论文解释了 AI 智能体如何通过观察屏幕而不是事先了解游戏的相关信息进行学习，就准确率而言，实验结果非常好。通过深度学习和强化学习的结合，开启了深度强化学习的时代。

Q-Learning 算法中的 Q 函数，能够根据当前状态估算奖励值，一般称之为 $Q(s, a)$，其中 Q 是一个函数，用于计算状态 s 和动作 a 的预期值。而在 DQN 算法中，则使用神经网络来基于状态估算奖励值，在第 10.2 节将详细讨论它的工作原理。

10.2 使用 Keras 进行《CartPole》游戏

《CartPole》是 OpenAI Gym（游戏模拟器）中一种简单的操作环境。其目标是平衡与移动车顶部的一个关节连接的杆，游戏并不给出像素信息而是给出两种状态信息：杆的角度和车的位置。智能体可以通过对小车执行 0 或 1 的一系列操作来移动小车，使其向左或向右移动：

OpenAI Gym 使得与游戏环境间的交互非常简单：

    next_state, reward, done, info = env.step(action)

在上面代码中，操作可以是 0 或 1。输送这些数字后，`env`（即游戏环境）将输出结果。`done` 变量是一个布尔值，表示游戏是否结束，旧的状态与 `action`、`next_state` 匹配，而 `reward` 是训练智能体所需信息。

怎么做

使用神经网络来构建 AI 智能体从而进行《CartPole》游戏。该网络包括含四个参数的输入层、三个隐藏层以及含两个可能的输出（0 或 1）的输出层。

1. Keras 使得基本神经网络的实现变得简单，用以下代码创建一个空的序贯模型：

```
model = Sequential()
model.add(Dense(24, input_dim=self.state_size, activation='relu'))
model.add(Dense(24, activation='relu'))
model.add(Dense(self.action_size, activation='linear'))
model.compile(loss='mse',
              optimizer=Adam(lr=self.learning_rate))
```

使用线性激活函数、**均方误差（MSE）**损失函数和 Adam 优化器来构建神经网络。

2. 对于基于环境数据进行预测的神经网络，必须馈入信息。`fit()`方法将输入和输出成对馈入模型，模型根据馈入的数据进行训练，从而基于输入来近似输出。

该训练过程使神经网络能够预测某一特定状态的奖励值：

```
model.fit(state, reward_value, epochs=1, verbose=0)
```

3. 训练之后，模型可以预测未知输入的输出。在模型上调用`predict()`函数，模型将根据训练数据来预测当前状态的奖励：

```
prediction = model.predict(state)
```

实现 DQN 智能体

在游戏中，奖励与游戏的得分成正比。在《CartPole》游戏中，当杆向右倾斜时，将按钮向右推的奖励将高于向左推，杆将保持垂直更长时间。为了清晰地表达这种逻辑并对其进行训练，必须将其公式化并进行优化，其中损失代表预测值与实际目标值之间的差异。

《CartPole》的损失公式如下所示：

$$loss = (r + \gamma * \max_{a'} Q(s', a') - Q(s, a))^2$$

其中：

- r：奖励
- γ：衰减率

- s：序列/状态
- a：操作
- a'：可能的操作
- s'：可能的下一个状态
- Q：最优操作值函数 $Q(s, a)$，即在输入某个序列 s，然后执行某个操作 a 后达到最大的预期回报。

使用 Keras 实现这一任务。通过一行 python 程序来定义预期目标：

```
target = (reward + self.gamma *
         np.amax(self.model.predict(next_state)[0]))
```

上述公式用于从预测输出中减去目标值并求其平方，这一公式还应用了创建神经网络模型时定义的学习速率，这个计算发生在 `fit()` 函数中。该函数通过学习速率，利用 Adam 的损失函数梯度下降减小预测值和目标值之间的差距。通过程序重复更新过程，Q 值近似收敛到实际 Q 值，损失减少，目标值得以改善并变得更准确。

DQN 算法两个最显著的特征是 `remember` 和 `replay`。两者都是很简单的概念，将在下面进行解释。

记忆与 remember

DQN 需要克服的挑战之一是算法中的神经网络往往会用新的经验覆盖旧的经验。但实际运行中需要先前的经验和观察清单，以便根据先前的经验重新训练模型，这些经验被称为**记忆**（memory），可以使用 `remember()` 函数将 `state`、`action`、`reward` 和 `next_state` 附加到记忆中。

`memory` 列表的形式如下所示：

```
memory = [(state, action, reward, next_state, done)...]
```

`remember` 函数将状态、操作和结果奖励值存储到记忆中，代码如下所示：

```
def remember(self, state, action, reward, next_state, done):
    self.memory.append((state, action, reward, next_state, done))
```

replay 函数

`replay()` 是一种利用记忆中的经验训练神经网络的方法。

1. 首先，从记忆中获取一些经验，将其称为 `minibatch`：

```
minibatch = random.sample(self.memory, batch_size)
```

2. 上述代码生成 `minibatch` 的过程为：随机从记忆中采样 `batch_size` 大小的元素，此例中 `batch_size` 大小为 32。

3. 为了使智能体长期发挥良好的作用，不仅需要考虑直接的奖励值，还要考虑未来的奖励值。为此，需要设置 `discount rate` 或 `gamma`，智能体将根据给定状态学会如何最大化未来的奖励值：

```
def replay(self, batch_size):
    minibatch = random.sample(self.memory, batch_size)
    for state, action, reward, next_state, done in minibatch:
        target = reward
        if not done:
            target = (reward + self.gamma *
                      np.amax(self.model.predict(next_state)[0]))
        target_f = self.model.predict(state)
        target_f[0][action] = target
        self.model.fit(state, target_f, epochs=1, verbose=0)
    if self.epsilon > self.epsilon_min:
        self.epsilon *= self.epsilon_decay
```

act 函数

智能体首先按特定百分比随机选择一个操作，这个百分比即 `exploration rate` 或 `epsilon`。首先，智能体在开始学习模式之前会进行各种尝试。随后，智能体基于当前状态预测奖励值并选择将会赋予最高奖励的操作。`np.argmax()` 用于在 `act_values[0]` 中选择两个元素之间的最大值：

```
def act(self, state):
    if np.random.rand() <= self.epsilon:
        return random.randrange(self.action_size)
    act_values = self.model.predict(state)
    return np.argmax(act_values[0])  # returns action
```

`act_values[0]` 的形式为：`[14.145181,11.2012205]`，其中数字分别代表操作 0 和 1 的奖励值。`argmax` 函数选择具有最高值的索引。对于示例中的 `[14.145181, 11.2012205]`，`argmax` 返回 0，因为 0 对应的索引值是最高的。

DQN 的超参数

如下为传递给 DQN 智能体的超参数：

- `episodes`：智能体将运行的游戏次数。
- `gamma`：衰减率或折扣率，计算未来的奖励值。
- `epsilon`：搜索率，智能体随机决定操作而不是预测操作的比率。
- `epsilon_decay`：表示游戏过程中劣势搜索次数的参数。
- `epsilon_min`：智能体至少搜索的次数。
- `learning_rate`：确定神经网络在每次迭代中学习的程度。

DQNAgent 类

`DQNAgent` 类具有已讨论过的 `embody`、`model`、`remember` 等方法：

```
class DqnAgent:
    def __init__(self, state_size, action_size):
    def _build_model(self):

    def remember(self, state, action, reward, next_state, done):
    def act(self, state):

    def replay(self, batch_size):

    def load(self, name):

    def save(self, name):
```

每种方法都包括以下函数：

- `__init__(self,state_size,action_size)`：用 `state_size` 和 `action_size` 参数初始化类：
  - state_size = 4
  - action_size = 2

- `_build_model(self)`：使用 Keras 序贯模型构建并返回神经网络模型，该模型有两个隐藏层，每层 24 个神经元。本例中输出层的输出 `action_size=2`。

- `act(state)`：基于已知状态来进行操作并预测奖励值。

- `replay(self,batch_size)`：利用记忆中的经验训练神经网络。

- `load(self,name)`：通过给定名称加载模型权重。

- `save(self,name)`：通过给定名称保存模型权重：

```
class DqnAgent:
    def __init__(self, state_size, action_size):
```

```python
        self.state_size = state_size
        self.action_size = action_size
        self.memory = deque(maxlen=2000)
        self.gamma = 0.95    # discount rate
        self.epsilon = 1.0   # exploration rate
        self.epsilon_min = 0.01
        self.epsilon_decay = 0.995
        self.learning_rate = 0.001
        self.model = self._build_model()

    def _build_model(self):
        # Neural Net for Deep-Q learning Model
        model = Sequential()
        model.add(Dense(24, input_dim=self.state_size, activation='relu'))
        model.add(Dense(24, activation='relu'))
        model.add(Dense(self.action_size, activation='linear'))
        model.compile(loss='mse',
                      optimizer=Adam(lr=self.learning_rate))
        return model

    def remember(self, state, action, reward, next_state, done):
        self.memory.append((state, action, reward, next_state, done))

    def act(self, state):
        if np.random.rand() <= self.epsilon:
            return random.randrange(self.action_size)
        act_values = self.model.predict(state)
        return np.argmax(act_values[0])  # returns action

    def replay(self, batch_size):
        minibatch = random.sample(self.memory, batch_size)
        for state, action, reward, next_state, done in minibatch:
            target = reward
            if not done:
                target = (reward + self.gamma *
np.amax(self.model.predict(next_state)[0]))
            target_f = self.model.predict(state)
            target_f[0][action] = target
            self.model.fit(state, target_f, epochs=1, verbose=0)
        if self.epsilon > self.epsilon_min:
            self.epsilon *= self.epsilon_decay

    def load(self, name):
        self.model.load_weights(name)

    def save(self, name):
        self.model.save_weights(name)
```

下面将研究如何训练上述类创建的智能体。

训练智能体

本节中将介绍智能体如何针对 EPISODES 进行训练，提高奖励值并重新计算 epsilon：

```python
if __name__ == "__main__":
    env = gym.make('CartPole-v1')
    output_file = open("cartpole_v1_output.csv","w+")
    state_size = env.observation_space.shape[0]
    action_size = env.action_space.n
    agent = DqnAgent(state_size, action_size)
    # agent.load("./save/cartpole-dqn.h5")
    done = False
    batch_size = 32
    count = 0

    for e in range(EPISODES):
        state = env.reset()
        state = np.reshape(state, [1, state_size])
        for time in range(500):
            # env.render()
            action = agent.act(state)
            next_state, reward, done, _ = env.step(action)
            reward = reward if not done else -10
            next_state = np.reshape(next_state, [1, state_size])
            agent.remember(state, action, reward, next_state, done)
            state = next_state
            output = str(e) + ", " + str( time) + ", " + str(agent.epsilon) + "\n"
            output_file.write(output)
            output_file.flush()
            if done:
                print("episode: {}/{}, score: {}, e: {:.2}"
                      .format(e, EPISODES, time, agent.epsilon))
                break
            if len(agent.memory) > batch_size:
                agent.replay(batch_size)
        # if e % 10 == 0:
        #     agent.save("./save/cartpole-dqn.h5")
    output_file.close()
```

执行示例后，如下为 score 和 epsilon 的时序函数输出：

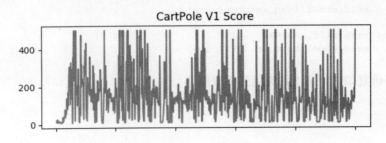

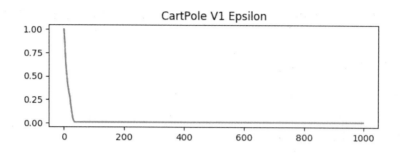

前面的一段时间内，得分差异很大，但随着智能体逐渐学习成为专业玩家，波动逐渐放缓。值得注意的是，当最初为随机值然后趋于稳定时，`epsilon` 也会大幅下降。

10.3 使用竞争 DQN 算法进行《CartPole》游戏

本节将介绍原始 DQN 网络的改进，**竞争 DQN 网络架构**（Dueling DQN Network）。它明确地分离了状态值的表示和（状态依赖的）操作优势。竞争架构由两个流组成，分别代表着共享着一个卷积特征学习模块的价值函数和优势函数。

通过聚合层将两个流组合并产生状态—操作函数 Q 的估计值，如下图所示：

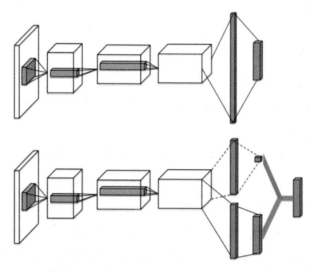

单流 Q 网络（上部）和竞争 Q 网络（下部）

竞争网络通过两个流来分别估计状态值（标量，用 $V(...)$ 表示）和每个操作的优势

（用 A（...）表示）。绿色输出模块通过以下方程将它们组合，两个网络都为每个操作输出 Q 值。

使用以下方程来代替定义 Q：

方程 a：

$$Q(s, a; \theta, \alpha, \beta) = V(s; \theta, \beta) + A(s, a; \theta, \alpha)$$

从优势函数中减去一项：$A(s, a; \theta, \alpha)$

$$\frac{1}{|A|} \sum_{a'} A(s, a'; \theta, \alpha))$$

一方面，由于除以一个常数，上式失去了 V 和 A 最初的意义。但另一方面，它增加了优化的稳定性：使用（）中的部分，使得优势函数像平均值一样快地变化，而不需要补偿方程 a 的最优操作对优势函数产生的变化。

方程 b：

$$Q(s, a; \theta, \alpha, \beta) = V(s; \theta, \beta) + \left(A(s, a; \theta, \alpha) - \frac{1}{|A|} \sum_{a'} A(s, a'; \theta, \alpha) \right)$$

其中

- s：序列 / 状态。
- a：操作。
- a'：可能的操作。
- s'：可能的下一个状态。
- Q：最优操作值函数 $Q(s, \alpha)$，即在输入某个序列 s，然后执行某个操作 α 后达到的最大预期回报。
- 一个流输出：全连接层标量输出，$V(s; \theta, \beta)$ 也称为**状态网络**。
- 其他流输出：$|A|$ 维向量 $A(s, a; \theta, \alpha)$，也称为优势网络。
- θ 表示卷积层的参数：α 和 β 是两个全连接层的流的参数。

该竞争网络可以理解为双流 Q 网络，替代之前算法中的单流 Q 网络，例如 DQN。竞争网络自动生成状态值函数和优势函数的单独估计，无须任何额外监督：

- https://www.cs.toronto.edu/~vmnih/docs/dqn.pdf
- http://proceedings.mlr.press/v48/wangf16.pdf

接下来介绍《CartPole》游戏中的实际竞争网络。

10.3.1 准备工作

使用 DQN 的 `keras-rl` 实现（https://github.com/keras-rl/keras-rl）：

```
import numpy as np
import gym
from keras.models import Sequential
from keras.layers import Dense, Activation, Flatten
from keras.optimizers import Adam

from rl.agents.dqn import DQNAgent
from rl.policy import BoltzmannQPolicy
from rl.memory import SequentialMemory
```

接下来介绍从 `keras-rl` 导入的三个类。

DQNAgent

这个类是 `keras-rl` 代码库的一部分，不需要单独配置，如下为具体实现部分：

```
class DQNAgent(AbstractDQNAgent):
    # class methods and body
```

init 方法

`init` 方法中用到如下参数：

- `model`：Keras 模型。
- `policy`：在规则（https://github.com/keras-rl/keras-rl/blob/master/rl/policy.py）中定义的 `keras-rl` 规则。
- `test_policy`：一个 `keras-rl` 规则。
- `enable_double_dqn`：一个使目标网络成为第二个网络，用于减小过拟合的布尔值，由 van Hasselt 等人提出。
- `enable_dueling_dqn`：Mnih 等人提出，用于支持竞争架构的布尔值。
- `dueling_type`：如果 `enable_dueling_dqn` 设置为 `True`，则必须选择一种竞争架构，用于分别从 $V(s)$ 和 $A(s, a)$ 计算 $Q(s, a)$。注意在论文中推荐 avg（https://arxiv.org/abs/1511.06581）。

```
avg: Q(s,a;theta) = V(s;theta) + (A(s,a;theta)-
Avg_a(A(s,a;theta)))
```

```
max: Q(s,a;theta) = V(s;theta) + (A(s,a;theta)-
max_a(A(s,a;theta)))
naive: Q(s,a;theta) = V(s;theta) + A(s,a;theta).
```

实现过程可参阅如下论文：

- **Mnih et al**: https://arxiv.org/pdf/1412.7755.pdf
- **Zing Wang**: https://arxiv.org/abs/1511.06581

设置网络的最后一层

基于已选的竞争类型，选择网络的最后一层并将其传递给 `init` 函数。例如，通过如下代码设置 avg 竞争类型的输出层：

```
if self.dueling_type == 'avg':
        outputlayer = Lambda(lambda a: K.expand_dims(a[:, 0], -1) + a[:, 1:] - K.mean(a[:, 1:],
                    keepdims=True), output_shape=(nb_action,))(y)
```

竞争规则

各类竞争规则：

- **Eps 贪婪规则**：要么以概率 epsilon 采取随机操作，要么以概率（1-epsilon）采用当前最佳操作。
- **softmax 规则**：根据概率分布采取操作。
- **线性锻炼规则**：计算当前阈值并将其传递给内部规则，内部规则负责选择当前操作，其中阈值会随着时间按照线性函数逐渐减小。

init 代码

上面描述的代码包含在以下代码段中：

```
class DQNAgent(AbstractDQNAgent):
    def __init__(self, model, policy=None, test_policy=None,
enable_double_dqn=False,
                enable_dueling_network=False,
                dueling_type='avg', *args, **kwargs):
        super(DQNAgent, self).__init__(*args, **kwargs)

        # Validate (important) input.
        if hasattr(model.output, '__len__') and len(model.output) > 1:
            raise ValueError('Model "{}" has more than one output. DQN expects a model that has a single output.'.format(model))
```

```python
        if model.output._keras_shape != (None, self.nb_actions):
            raise ValueError('Model output "{}" has invalid shape. DQN expects a model that has one dimension for each action, in this case {}.'.format(model.output, self.nb_actions))

        # Parameters.
        self.enable_double_dqn = enable_double_dqn
        self.enable_dueling_network = enable_dueling_network
        self.dueling_type = dueling_type
        if self.enable_dueling_network:
            # get the second last layer of the model, abandon the last layer
            layer = model.layers[-2]
            nb_action = model.output._keras_shape[-1]
            # layer y has a shape (nb_action+1,)
            # y[:,0] represents V(s;theta)
            # y[:,1:] represents A(s,a;theta)
            y = Dense(nb_action + 1, activation='linear')(layer.output)
            # caculate the Q(s,a;theta)
            # dueling_type == 'avg'
            # Q(s,a;theta) = V(s;theta) + (A(s,a;theta)-Avg_a(A(s,a;theta)))
            # dueling_type == 'max'
            # Q(s,a;theta) = V(s;theta) + (A(s,a;theta)-max_a(A(s,a;theta)))
            # dueling_type == 'naive'
            # Q(s,a;theta) = V(s;theta) + A(s,a;theta)
            if self.dueling_type == 'avg':
                outputlayer = Lambda(lambda a: K.expand_dims(a[:, 0], -1) + a[:, 1:] - K.mean(a[:, 1:], keepdims=True), output_shape=(nb_action,))(y)
            elif self.dueling_type == 'max':
                outputlayer = Lambda(lambda a: K.expand_dims(a[:, 0], -1) + a[:, 1:] - K.max(a[:, 1:], keepdims=True), output_shape=(nb_action,))(y)
            elif self.dueling_type == 'naive':
                outputlayer = Lambda(lambda a: K.expand_dims(a[:, 0], -1) + a[:, 1:], output_shape=(nb_action,))(y)
            else:
                assert False, "dueling_type must be one of {'avg','max','naive'}"

            model = Model(inputs=model.input, outputs=outputlayer)

        # Related objects.
        self.model = model
        if policy is None:
            policy = EpsGreedyQPolicy()
        if test_policy is None:
            test_policy = GreedyQPolicy()
        self.policy = policy
        self.test_policy = test_policy

        # State.
        self.reset_states()
```

玻尔兹曼 Q 规则

在探索中，希望将现有网络的 Q 估计值的所有信息利用起来。玻尔兹曼做到了这一点。这种方法并不总是采取随机操作或最优操作，而是选择具有加权概率的操作。为了实现这一点，它在每个操作的网络估计值上使用了 softmax，使得智能体估计最有可能（但不保证）选择最优操作。与 e-greedy 算法比起来，它最大的优势在于可以兼顾其他相关操作的值信息。如果智能体有四种可用的操作，那么在 e-greedy 中，三种非最优操作的估计值被认为是相同的，但在玻尔兹曼探索中，它们被根据相关值赋予不同权重。通过这种方法，智能体可以忽略被估计为次优的操作，而更多地关注那些有潜力但不一定理想的操作。

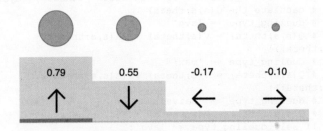

上图中，每个值对应于环境中随机状态（s）下给定操作（a）时的 Q 值，图中浅蓝条的条对应于选择给定操作的概率，深蓝条对应于当前选定的操作。

训练过程中的调整

实际中，一般利用额外的温度参数（τ），该参数随着时间减小。它控制 softmax 分布的传播，以便在训练开始时平等地考虑所有操作，并在训练结束时将操作稀疏分布。

在数学术语中，该规则可以描述为如下公式：

$$p_t a = \frac{\exp\left(\dfrac{q_t(a)}{\tau}\right)}{\sum_{i=1}^{n} \dfrac{q_t(i)}{\tau}}$$

该规则的初始化如下所示：

```
class BoltzmannQPolicy(Policy):
    """Implement the Boltzmann Q Policy
    """
```

```
def __init__(self, tau=1., clip=(-500., 500.)):
    super(BoltzmannQPolicy, self).__init__()
    self.tau = tau
    self.clip = clip
```

玻尔兹曼规则使用上面的公式进行定义，而具体操作时则使用如下代码实现：

```
p = exp(Q/tau) / sum(Q[a]/tau)
```

下面介绍如何存储这些奖励、操作和状态。

顺序存储器

DQN 智能体使用顺序存储器来存储各种状态、操作和奖励，具有以下数据结构：

- 观察（字典型）：观察环境并返回的观察结果。
- 操作（整型）：为获得观察值而采取的操作。
- 奖励（浮点型）：通过操作而得到的奖励值。
- 终端（布尔型）：状态终端。

上述数据结构的代码如下所示：

```
self.actions = RingBuffer(limit)
self.rewards = RingBuffer(limit)
self.terminals = RingBuffer(limit)
self.observations = RingBuffer(limit)
```

10.3.2 怎么做

通过以下步骤实现基于竞争 DQN 智能体的《CartPole》游戏程序：

1. 初始化 Open AI gym 环境 `env`。

2. 定义 `env` 中的操作数。

3. 创建一个序列神经网络。

4. 初始化 `SequentualMemory`，其中 `limit` 设置为 100，`window_length` 设置为 1。

5. 初始化 `BoltzmannQPolicy` 实例规则。

6. 创建 `DQNAgent`，如下所示：

```
dqn = DQNAgent(model=model, nb_actions=nb_actions, memory=memory,
nb_steps_warmup=10,
```

```
            enable_dueling_network=True,
target_model_update=1e-2, policy=policy)
```

7. 编译 `DQNAgent`，优化方法为 Adam，损失函数为**平均绝对误差（MAE）**。

8. 调用 `dqn.fit` 获得奖励值：

```
ENV_NAME = 'CartPole-v0'

# Get the environment and extract the number of actions. initiate
seed.
env = gym.make(ENV_NAME)
np.random.seed(123)
env.seed(123)
nb_actions = env.action_space.n

# A simple model regardless of the dueling architecture
# if you enable dueling network in DQN, DQN will build a dueling
network base on your model automatically

model = Sequential()
model.add(Flatten(input_shape=(1,) + env.observation_space.shape))
model.add(Dense(16))
model.add(Activation('relu'))
model.add(Dense(16))
model.add(Activation('relu'))
model.add(Dense(16))
model.add(Activation('relu'))
model.add(Dense(nb_actions, activation='linear'))
print(model.summary())
# Configure and compile the agent.
memory = SequentialMemory(limit=100, window_length=1)
policy = BoltzmannQPolicy()
# enable the dueling network
# dueling_type defined as {'max'}
dqn = DQNAgent(model=model, nb_actions=nb_actions, memory=memory,
nb_steps_warmup=10,
               enable_dueling_network=True,
target_model_update=1e-2, policy=policy)
dqn.compile(Adam(lr=1e-3), metrics=['mae'])

# now train the model
dqn.fit(env, nb_steps=1000, visualize=False, verbose=2)
```

通过修改 DQN 智能体将训练的结果 `metrpet:ics` 存储在文件中，当运行时，生成 `log` 文件，如下所示：

```
20,0.46619212958547807,0.5126896699269613,-0.0582879309852918
73,0.34128815,0.47138393,0.13162555
106,0.13377163,0.5142502,0.5694133
116,0.041925266,0.5914798,0.962825
139,0.016332645,0.6163821,1.1230623
```

```
198,0.0054321177,0.7427029,1.4339473
225,0.008107586,0.8838398,1.7260511
237,0.009388918,0.9404013,1.8322802
...
1000,0.1786698,4.1688085,8.251708
```

行的标题是 `episode`、`reward` 和 `steps`。

训练和测试结果绘图

对 `output` 文件夹中 `log` 文件的训练结果绘图，如下图可见，损失很快得到明显下降，但最终上升为异常值并再次降低。

MAE 和 `mean_q` 随着 `episodes` 上升而稳定上升。

1. `mean_q` 由以下函数定义：

```
def mean_q(y_true, y_pred):
    return K.mean(K.max(y_pred, axis=-1))
```

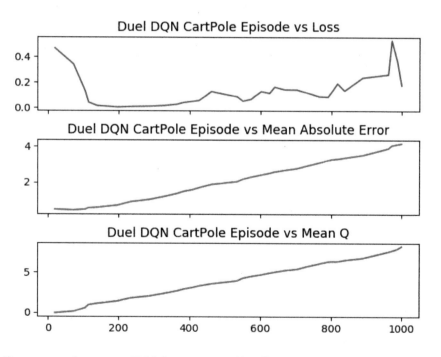

2. 将 `reward` 和 `steps` 绘制为 `episode` 的函数：

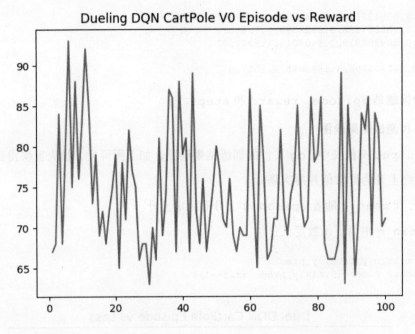

可以看出在 100 个 episode 中，reward 在 65 ~ 95 之间振荡。